TENSOR TECHNIQUES IN SYNMECHANICS

A new statistical dialect for multiscale analysis of plant particle chronoseres

I0481002

Back cover photo. I am standing on the bank of the Uruguay river in Parque do Turvo, a major conservation area in Rio Grande do Sul where the elements of several floras find refugia on the highly diverse terrain. Behind me, the turbulent flow in the lower channel on the Brazilian side. Beyond, the long waterfall called Salto do Yucumã on the Argentine side. Overflow from the river's upper channel creates the fall. The picture was taken on a field trip with students and faculty from UFRGS of Porto Alegre (April 27, 2003).

Banner headline. Conceptual tools are presented for Multiscale Time Series Analysis and tested on a long chronosere of plant particle counts taken from lake sediment in Eastern Beringia's dry arctic steppe tundra. The analysis identifies statistically significant directional regularities in the plant particle chronosere from which assembly rules are inferred that control the composition of the attractor (climax) state the source vegetation. The rules facilitate the construction of predictive assembly scenarios under the assumed change in the global climate.

Another project is done. What next for me at 88?
Perhaps, time for retirement with darling Márta.

On Molokai Princes leaving a safe harbour in Lahaina, Maui.

TENSOR TECHNIQUES IN SYNMECHANICS

A new statistical dialect for multiscale analysis of plant particle chronoseres

László Orlóci

BSF, DFE, MSC, PHD, DSC, FRSC, MHAS
INTECOL Distinguished Statistical Ecologist; Professor and Visiting Professor at Universities in the Americas, the Pacific, Asia, and Europe; Emeritus Professor of Statistical Ecology, Western University, London, Ontario, Canada

SCADA Publishing – 2020

Refer to this book:

Orlóci, L. 2020. Tensor techniques in synmechanics. A new statistical dialect for multiscale analysis of plant particle chronoseres. SCADA Publishing, Canada.
https://www.amazon.ca/dp/B08JGDYCX9

ISBN: 9798688389579
Imprint: Independently published

Tensor techniques in synmechanics

All rights reserved 2020 © László & Márta Orlóci
Direct all correspondence to lorloci@uwo.ca

Table of Contents

Foreword 5
Site and data 6
    Hanging Lake 6
    Three chronoseres 7
Palynomorph taxonomy 10
Frame..11
Tensors in synmechanics 13
Probing for Markov-type regularity..16
    Process regularity...16
    Looking back..17
    Stress-based statistical significance 19
    Statistical significance testing 21
    Example..22
    Climax prediction 26
Comparison of Vostok and Renland 39
Synmechanics 47
    Chronosere tensors...47
    Tensor vector analysis..48
Generalizations 55
Literature 55
Appendices 58

Foreword

I revisit Professor L.C. Cwynar's (1982) plant particle chronosere from Hanging Lake in the east Beringian arid steppe tundra – now the third time my trusted resource in essays on the vegetation process. The essays develop conceptual tools, sensitive at scales of resolution needed to lay open historic directional regularities with evidence entombed in deep lake sediments. On the first visit, a long decade ago, I probed for long-term patterns in moment and product moment-type parameter oscillations. I presented results (Orlóci 2009, 2012, 2019a) and named my approach "Multiscale Trajectory Analysis". When I returned on the second visit I searched for long-term energy-based entropy oscillations

(Orlóci 2014a,c, 2019a). At that time "Statistical Quantum Ecology" was born.[1] I offered it as a holistic alternative in the study of plant community energetics. With those techniques, I could go further, wider, and deeper with laying open the long-term process in the source vegetation than I possibly could straight-jacketed by the traditional techniques. In the present essay, I start from a rather different conceptual base, my forest engineering background. I am presenting "Multiscale statistical synmechanics", my tag for plant community mechanics. The major tools include the tensor vector concept, Markov chain mathematics, integral and differential calculus, and Monte Carlo simulation. I approach the task as a unique opportunity to lay foundations for yet a third statistical dialect (Orlóci 2015) with tensor-based stress as my pair-function.

As my first task following the acknowledgements, I introduce the Hanging Lake site and presented the data sets before offering thoughts on an operative taxonomy, the conceptual frame and techniques.

Acknowledgements. I am very much aware I use borrowed public domain data. I salute Professor L.C. Cwynar for the records, collected with such care and high professionality at the expense of so much personal effort, time and money, under the harshest field conditions imaginable, made available for the public domain. I can imagine the intent, to inspire in others research interest in probing the data further for insights waiting to be discovered. I received editorial help from Ms Martha B. Orlóci (BA Hon). Professor M. Mukkattu suggested structural changes and pointed out stylistic inadequacies. I extend my sincerest thanks to both for their help.

Site and data

Hanging Lake

Hanging Lake is a small ancient water body situated in the north-western corner of the Yukon Territory in Canada. I use Hultén's 1937 term 'Beringia' to name the arid, cold steppe vegetation formation which

---

[1] "Quantum" refers to base in Max Planck's (1901) work on the distribution of entropy in the resonator complex. His entropy is energy based, i.e. the natural log of the inverse of the total count of energy quanta in the complex. The resonators are elementary particles. I adopted an analogue complex model for the plant community's mega sale naming taxa as the 'resonators' and using taxon performance units (density, frequency, or some other) for energy units. I note, entropy is a physical concept that applies through scales from nano to mega.

encloses the Hanging Lake site in the Arctic tundra. The formation extends westward through Alaska well-into Siberia. The geographer's use of Beringia identifies a geographic area without a definite biogeographic connotation. Map 1 shows more details about the Hanging Lake site.

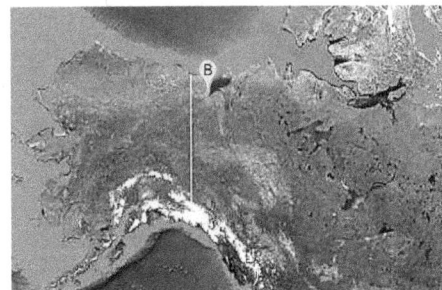 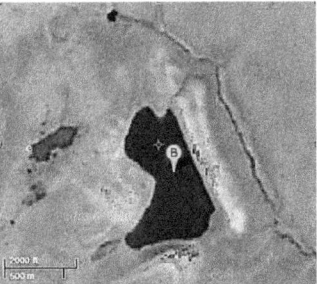

Map 1. Google Earth map showing the location of Hanging Lake (B), 68.355787 N, 138.359613 W, Yukon Territory, Canada.

Hanging Lake is situated in the foothills of the Barn Mountain on shale and sandstone terrain at about 530 m above sea level. The lake's surface area is roughly 60 ha and the water depth is less than 10 m. The site is located roughly 80 aerial kilometres due south from the Beaufort Sea and 110 km due east from the Alaska border. Note the sharp straight lines on the map. These are indicative of a relief unaffected by glaciation.

I should refer to Cwynar's (1982a,b) account for the full documentation. We learn from him, Hanging Lake has its drainage basin, no inlet stream, and no surface contamination, but the sediment core shows signs of solifluction. Like so many other lakes in the region, Hanging Lake has thermokarst origin.

Three chronoseres

I limit the analysis to just over 40000 sediment core years, long enough to capture major climatic events of the last glaciation and study their effect on the vegetation. Among the noted climatic events, there is the plummeting of the global atmospheric temperature to its lowest 24000 yrs. BP, the sharp temperature rise beginning 17000 yrs. BP and a maximum reached 8000 yrs. BP. These values are read directly from Figure 1a. The readings differ somewhat in Figure 1b. Temperature differences (dt) are shown involving two temperature readings, one based on the deuterium level measured in the ice core and the other based on the present deuterium level in the site.[2]

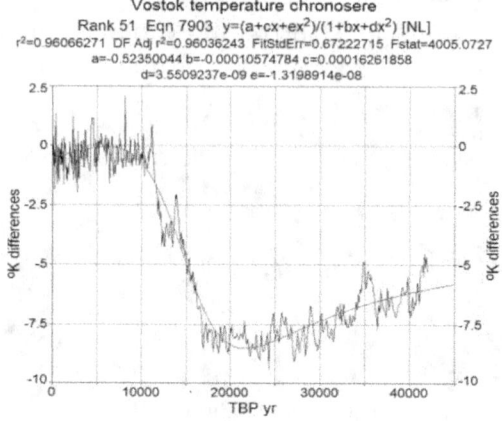

Figure 1a. Last 42k years in Petit's Vostok temperature chronosere from Antarctica.[3] Graphs: temperature observed, temperature predicted (trend line). The number of temperature measurements: 661. Core time interval: 0 to 41112 years. Ice core depth applied: 0 - 649 m. Regression analytical details[4] are in Appendix 1a. Vostok deuterium (D) difference to age difference (dt) transformation dt = 0.165(D_{sample} − $D_{present}$).[5]

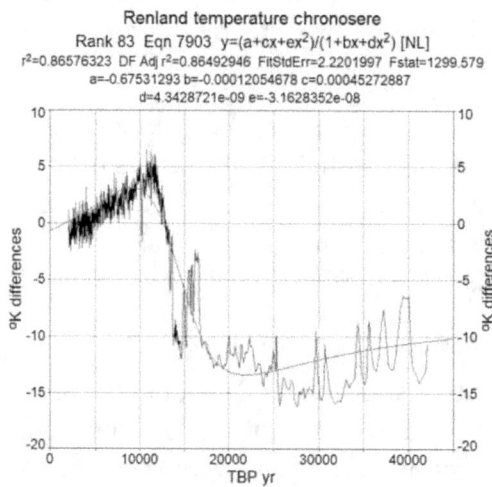

---

[2] The deuterium is extracted from the air bubbles in the ice. Deuterium originates in the inversion layer of the troposphere where the precipitation forms.

[3] Source address in 2010: www.ngdcnoaa.gov/paleoeo/icecore/antarctica/vostok

[4] Application TABLECURVE.

[5] Alternative offered in Schwinegruber (19960; T = (D+129.3)/5.666

Figure 1b. The Renland ice core temperature chronosere from Greenland.[6] Graphs: temperature observed, temperature predicted (trend line). The number of temperature measurements: 811. Core time interval: 2000[7] to 42070 years. Ice core depth applied: 0 – 312.66 m. Regression analytical details in the table in Appendix 1b. The Vostok records use different coefficients in the deuterium (D) difference to age difference (dT) transformation; my choice 0.33^{-1}, dt = (D_{sample} − $D_{present}$)/0.33.

With no denuding effects of magnitude, the land surface in the Hanging lake site retained conditions favourable for the preservation of deep, plant-particle bearing sediments. The sediment core encapsulates a plant particle spectrum captured by the Cwynar chronosere A partial spectrum of selected palynomorph taxa is displayed in Figure 2.

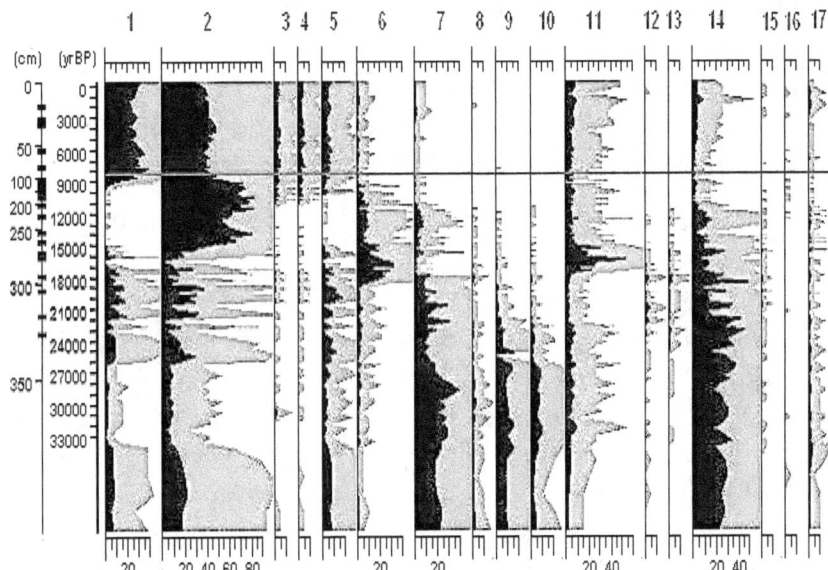

Figure 2. L.C. Cwynar's graph presenting a selection of the Hanging Lake plant particle chronosere. Columns 1 to 17 show particle counts for individual palynomorph taxa and two groups: 1 Alnus, 2 Betula, 3 Ericaceae, 4 Ericales[8], 5 Picea, 6 Salix, 7 Artemisia, 8 Asteraceae/Asterideae, 9 Brassicaceae, 10 Chenopodiaceae/Amaranthaceae, 11 Cyperaceae, 12 Fabaceae, 13 Plantago canescens, 14 Poaceae, 15 Rosaceae, 16 Unspecified groups of trees and shrubs, 17 Unspecified herbs. Bottom scale: % count. Dark shading: original scale. Light shading: 5x original scale. Black markings on depth scale: carbon dated horizons. The bottom layer in sediment at 398 cm dated (by

---

[6] http://www.tellusb.net/index.php/tellusb/article/view/1581temperature

[7] A more recent data set combines data from two ice cores to fill the empty 0 to 2000 yr periods in the data (Vinther et al. 2008).

[8] Not an error. Ericaceae and Ericales are taken by Cwynar as different palynomorph taxa. See next chapter.

extrapolation) to 41138 yr BP. The horizontal line at the 8299-year mark: a recordset, one of the 133 paleorelevés. Source database identified in footnote 9.

Figure 2 portrays 15 of Hanging Lakes palynomorph taxa and 2 groups. The chronosere I downloaded from the NOAA database gives particle density counts of 89 taxa in 133 sediment horizons. As I am writing these, the University of Ottawa address[9] is still functional for obtaining the 89x133 data set.

The spectrum as presented in Figure 2 has mutilated in several respects. The downloaded data set is complete. A report attached to the NOAA data set gives details of the then-current vegetation of the Hanging Lake site on the specified date. A total of 47 species of vascular plants, 12 bryophytes, and 18 lichens are identified locally within tundra types. The types are named by the vegetation's physiognomy (tussock, heath) and the quality of the substrate (shale felt fields, and sandstone slope).

I used the term palynomorph taxa and now I should explain.

Palynomorph taxonomy

The initial task in the construction of any plant particle chronosere is the identification of fossilised pollen grains, plant spores, algal cells, and other plant tissues extracted from the sediment. Identification is the way to assign particles carrying specific sets of key traits to 'particle types' called 'palynomorph taxa'. The taxa in the same sediment horizon are elements of the horizon's paleo collection. Such a community is not synbiological. Its source for particles is. The paleo community is mechanical and has virtual existence only as a chance-based plant particle assemblage, a sample of the limitless metacommunity from which its members originate. Dispersion is involved that brings the members into the same sediment layer. In this regard, it may be called the source metacommunity's diaspora whose natural concentric belts extend progressively further by accessibility from the sediment core's site. The members in successive belts have a diminishing chance of being represented in the sediment core sample as the distance increases. Air

---

[9] Cwynar 1982 – http://www.lpc.uottawa.ca/data/cpd/HangingLake.xls
Global Pollen Database (2000) http://www.ncdc.noaa.gov/paleo/ftp-pollen.html
http://hurricane.ncdc.noaa.gov/pls/paleox/f?p=519:1:0::::P1_STUDY_ID:7483

currents, flowing water, gravity, and the fauna are the selecting and transporting agents. The vector description of a paleo community **X** within a sediment layer is my paleorelevé. Particle counts in **X** are the relevé's Cartesian co-ordinates.

It is important to realise that the physical state of a sediment-born plant particle is rather poor and plant identification based on the usual taxonomic criteria applied in modern phylogenetic practice bounds to meet with only limited success. For example, we already encountered two palynomorph taxa Ericaceae and Ericales. The presence of both in Cwynar's shortlist (Figure 2) is not an error. It is just a consequence of the taxonomist not having been able to identify the plant particles to a lower level of plant systematics in that specific case. Ericales and Ericaceae are accepted as distinct palynomorph taxa. I warn students, the use of the standard taxonomic nomenclature should not mislead them to expect species-level phylogenetic homogeneity in a palynomorph taxon. Palynomorph taxa are not species, not genera, not families, and not anything other than arbitrary categories justified by the method of identification. But palynomorph taxa are distinct paleo vegetation types aggregating distinct populations of plant particles. In such populations, a mixed inheritance of the parental sources is the rule, not the exception. I discuss relevant topics in an earlier essay (Orlóci 1991).

Frame

The frame is a conceptual map of the analysis. In our case, it is drawn for laying open the historic regularities in the plant particle chronosere concomitant with historic temperature oscillations in the atmosphere. We need two types of data. Cwynar's chronosere gives us the vegetation's community-level assembly details. Records of long-term temperature oscillations are available from the Arctic and Antarctic ice cores. I have interesting statistical results to present about the functional synchrony of temperature oscillation from the two sources and the tensor states of the plant particle chronosere in East Beringia's dry steppe tundra.

I must say, the Cwynar plant particle chronosere is a textbook example of vegetation time series born in Nature's chaos[10]. I explain in the sequel how did this fact influence my choice of the frame within which I select the data sets, state the objectives, do the analyses, and argue why I am opting to create a new methodology, which I already named multiscale statistical synmechanics.

In the discipline I am defining the frame, it is commonplace to speak of holism and reductionism, and to present them as alternatives. I do not see them that way, but rather as a dichotomy opening the choice of one or the other as an answer to a simple question: which of the two has the potential of returning maximum information useful to me. I note when emergent properties are involved the choice is clear: holism.

Gleick's (1988) book "Chaos" has much on the topic. Arguments are marshalled why modern science has returned to holism to solve hitherto unsolvable problems related to complex systems. I consider the vegetation process such. I find Gleick's passage my favourite in which Gleick discusses an example from Gottfried Wilhelm Leibniz. Extemporaneously quoted, it goes like this:

Leibniz imagines ... a drop of water contained a whole teeming universe, containing, in turn, water drops and new universes within.

The water drops' self-similarity is obvious. One drop is like any other, or it seems that way. What does this tell us? The small water drop is a scaled-down version of the large water drop, but not implying the small drop could reveal the contents of the large drop. In other words, reductionism has a limit to characterise a complex system.

By extension of Leibniz's idea, we may ask what connects the whole, in our case the vegetation community, and its elementary parts, the individual plants? The link must be the design that mitigates the community's assembly process. Interestingly, the design is not something that we expect to find on the first inspection of the system. When it is latent, intangible, it can be detected only from its manifestations, the emergent attributes. To identify the best design, we need to find the

---

[10] Interpret as a chaos theoretical term (Gleick 1987).

scale at which the manifestations are least fuzzy. For this, the approach must be multiscale.

When we take a plant particle chronosere, the particle collection by sediment horizon is proxy for the source vegetation's momentary composition. The elementary plant particles take the place of the individual plants. If I follow Polanyi's (1968) logic about natural systems, I should recognise two realities in my source vegetation. One is the community, the whole at the highest level, Leibniz's large water drop. The other is the individual plant, the indivisible element of the source vegetation, the lowest level of reduction, Leibnitz's smallest water drop. All intermediate reductions are arbitrary. The message is clear: the elementary level cannot specify the whole without knowing the assembly rules. Considering this, the controversy fanned by the band-wagon slogan in science, 'back to the parts' is vacuous.

I take this thought a step further when I say chronoseres of plant particle counts (densities) record information about two realities. The classical tools of palaeobotany take interest in the particle level and interpret vegetation history on that basis. There is not much left for me to do on that level without being slavishly repetitious. The collection level, the level of the paleorelevé or paleo community, is different. The manifestations of the assembly process come into full view on that level. Vivat collection! Hale paleorelevé!

Tensors in synmechanics

Tensors are directed attributes equipped to capture the magnitude and direction of change in a manner that can be analysed by the techniques of synmechanics. Process acceleration is an example. It is a directed attribute.

Tensors come in vectors; therefore, tensor vectors have direction. To see that direction matters, consider what we perceive as the level of acceleration when we view the process from one direction, we perceive as the level of deceleration from the opposite direction.

In applications, a case of the natural vegetation's climax process is my example. The natural climax, a virtual state in the future, a kind of moving target, an attractor as it were, may not even be reached. Yet, the process is following it seamlessly. Process trajectory is traceable from its footprints left in the past. The plant particle chronosere of Hanging Lake is such a footprint. I return to synmechanical tensors in detail after I present my reasoning regarding the choice of approach, the "holistic frame" in the analysis.

I am posing specific questions in syndynamics for which I hope to find answers in tensor vector analysis. The general description of tensors from the earlier sections is an adequate base for a slightly more technical definition. Among many I have seen a tensor vector circumscribed this way:

"... a vector of components (elements), functions of the coordinates of any point in geometric space, that transforms linearly between coordinate systems."

This definition is clear to me and serves my purpose, but it calls for explanation to clarify what makes such a definition usable in the analysis of a plant particle chronosere:

1. Each sampled sediment horizon is described by a recordset I call paleorelevé. The elementary records are plant particle counts of palynomorph taxa. The relevé collection in its natural order by sediment depth defines the trajectory of the vegetation process.

2. The tensor vector has elements in number equal to or less than the number of paleorelevés. Take two sediment horizons and label them j and k. Sediment layer j is closer to the bottom of the sediment; k is closer to the top. The direction from trajectory point j to k mark time forward in the vegetation process.

3. The tensor that defines the vector elements may identify a mechanical property such as compositional velocity $v_{j \to k}$ written for plant particle records of two sediment horizons, j and k, in the chronosere in the arrow's direction. This makes the tensor vector in which $v_{j \to k}$ is an element directed. As described, the vector is pointing time-forward to an attractor we call the climax (state) of the syndynamic process.

4. Given j and k, two paleorelevés are selected: $\mathbf{X}_j = (X_{1j}\ X_{2j} \dots X_{sj})$ and $\mathbf{X}_k = (X_{1k}\ X_{2k} \dots X_{sk})$. A characteristic element X_{1j} is the count of palynomorph taxon 1 in sediment horizon j and X_{1k} at sediment horizon k. The

vegetation process implies a change in composition, composition X_j giving way to composition X_k in situ, referencing the source vegetation, over time. Kerner von Marilaun has described the phenomenon in 1863, we attribute the process to the complex mechanism called facilitation.

5. Cwynar's records contain 133 paleorelevés. I denote this number by n. The set X_1, X_2, ..., X_{133} describe the plant particle chronosere. My symbol for the number of taxa is s, in value 89 for the 133 relevés in the 1982 records. Occasionally I refer to X_j and X_k as position vectors. Their tips are d_{jk} distance apart (1) in plant particle units.

$$d_{jk} = \left[\sum_{i=1}^{s} (X_{ij} - X_{ik})^2 \right]^{1/2} \quad j=1 \text{ to } n-1, \ k=j+1 \quad (1)$$

The sequentially defined set of distances $D = (d_{12} \ d_{23} \ d_{34} \ ... \ d_{132\,133})$ is a tensor vector too.[11] D is a proxy distance structure in the syndynamic of the source vegetation.

I introduced earlier the idea of synmechanical tensors. I use as such compositional distance d_{jk} taken from vegetation state X_j to the next state X_k time forward in the chronosere. Other synmechanic tensors include process velocity $V_k = \dfrac{d_{kj}}{t_k - t_j}$, $j \rightarrow k$ (2) and acceleration $A_k = \dfrac{V_k - V_j}{|t_k - t_j|}$, $j \rightarrow k$ (3). Interpret the arrow in $j \rightarrow k$ as a sign of process direction, such that relevé j closer to the sediment core's bottom is paired with its neighbour relevé k closer to the sediment core's top. Kinetic energy (linear momentum) is defined as $L=mV$ (4). In this m stands for mass $m = \sum_{i=1}^{s} X_{ij}$. Force is defined by $F=mA$ (7) and work by $W=Fs$ (5).

---

[11] D is a directed vector by virtue of its chronosere attachment. The d_{jk}'s tensor quality makes D a tensor vector. Note that d_{jk} satisfies the four metric properties (Orlóci 2014a, Section 8.2) and for that reason the D configuration's $n-1$ elements are invariant under any linear transformation. Eigenanalysis, the mathematical engine of principal components analysis (Orlóci 2019, Section 16.2), is an example of linear transformations.

V, A, L, F, and W are tensor vector elements. Their number is as many per vector as the number of distinct paired comparisons among the n relevés. In chaining, the number is n-1=132 and in full dimension n(n-1)/2=8778. Tensor vectors V, A, L, F, and W have direction, the same as the process direction, from past to present and are invariant under linear transformation of X.

Probing for Markov-type regularity

Process regularity

The first major question I pose regards detection of signs that indicate the presence of process regularity other than the random. True, there are many types of regularities and many could make sense to prob for in the chronosere, time and resources permitting. This not being the case, and a strong personal inclination to have more fun with the Markov chain explains why I have the contents as I did, in much of the monograph's following sections.

I experimented with Markov chains long enough not to expect more than a weak, but significant Markov manifestation. The main culprit for this is the immensity of the random effect which masks the effect of mechanisms responsible for directed regularity.

I reduce the task of detecting Markovity to a simple statistical problem: how similar two geometric (matrix algebraic) structures D and Δ must be before I can consider their similarity significant at given p probability. One of the structures is defined by the observed distance matrix D and the other by Δ, a simulated Markov distance matrix. My decision parameter for testing Markovity's significance is the engineer's stress function $\sigma(D,\Delta)$ (equation (2), Orlóci 2014 and references therein). $\sigma(D,\Delta)$ measures stress in D that would have to be overcome to transform it into the structure Δ. When D has $n-1$ distinct elements, d_{12} d_{23} ... $d_{n-1\,n}$, we use equation 2a. When D has $n(n-1)/2$ elements d_{12} d_{13} ... d_{1n} d_{23} d_{24} ... d_{2n} ... $d_{n-2\,n-1}$ $d_{n-1\,n}$, we use equation 2b.

$$\sigma^2(D,\Delta)=\frac{\sum\limits_{j=1,k=j+1}^{n-1}(d_{jk}-\delta_{jk})^2}{\sum\limits_{j=1,k=j+1}^{n-1}(d_{jk}+\delta_{jk})^2} \quad (2a) \qquad \sigma^2(D,\Delta)=\frac{\sum\limits_{j=1}^{n-1}\sum\limits_{k=J+1}^{n}(d_{jk}-\delta_{jk})^2}{\sum\limits_{j=1}^{n-1}\sum\limits_{k=J+1}^{n}(d_{jk}-\delta_{jk})^2} \qquad (6)$$

The elements δ_{kj} in $\mathbf{\Delta}$ are defined by the same function (1), except for **X** which in this case represent elements the Markov co-ordinates.

The exact form of $\mathbf{\Delta}$ depends on **P**, an $s \times s$ matrix of transition probabilities. This matrix is not symmetric, and the rows have unit sums. Any arbitrary set of $s \times s$ numbers whose rows are normalised will define a Markov chain. **P** may come from theory; I extract it from the observed chronosere **X** (1993; Orlóci 2019, Section 20.7). Any test of significance requires random expectations for $\sigma(\mathbf{D}, \mathbf{\Delta})$. Random expectations can be generated in Monte Carlo experiments. The test criterion is written the usual way. I will return to the topic when I use the test the first time.

$\sigma(\mathbf{D}, \mathbf{\Delta})$ varies within upper limit 1 and lower limit 0. $\sigma(\mathbf{D}, \mathbf{\Delta})$ is invariant under the linear transformation of **X**, and its one complement measure stress over the full length or a segment of the chronosere. When $\sigma(\mathbf{D}, \mathbf{\Delta})$ = 0, **D** and $\mathbf{\Delta}$ represent identical structures. Any value larger than 0 indicates the presence of stress. Stress is generated in the process of facilitation that resolves into a steady state as the attractor is approached if the attractor stays stationary. In phytosociological jargon, the climax state is the climax community ruled by chaos. I just repeat, chaos is not the same as pure randomness. A deluge of randomness is a part of the chaos.

Looking back

I am a believer of the classical phrase *Repetatio est mater studiorum*. So, let me repeat:

1. The sampling unit is the paleorclevé, an s-valued vector $\mathbf{X} = (X_1 + X_2 + ... + X_s)$. It is a set of s particle counts taken from a sediment core at a given depth. If I want to specify the horizon of origin, I add a subscript j such as $\mathbf{X}_j$. The elements of $\mathbf{X}_j$ carry the subscript in the manner of X_{ij}. I use the X_{ij} as co-ordinates of an s-dimensional Euclidean space. In that space, the tip of **X** is the point image of a paleorelevé. If there are two paleorelevés, $\mathbf{X}_j$ and $\mathbf{X}_k$, they define two points whose distance is d_{jk}. By convention, relevé j is taken to be closer than k to the sediment's bottom.

2. **D** is a distance vector. The elements represent the distances of neighbouring paleorelevés in the chronosere in order starting with initiation, usually the bottom horizon in the sediment core and moving up to present, usually the top sediment horizon.

3. **D** can have less than n-1 or n(n-1)/2 values, the maxima of the two alternatives we defined.

4. **D** is a tenser vector, a proxy carrier of information about syndynamics in the source metacommunity.

5. **D** does encapsulate all effects under which the paleo particle spectrum came into existence.

6. When we probe **D** for signs of directed regularity, we do it comparatively under the hypothesis that directedness has a specific mathematical form other than random. At this time, I choose to examine the Markov chain.

7. Regarding the vegetation process, its mechanics fixed in the past, is indeed pointing into the future. The process has been modelled by the Markov chain[12] and on that basis the future process states predicted are Markovian.

8. Markovity is statistically testable based on the $\sigma(D, \Delta)$ distribution at different orders. The test is conditioned by the fact that Markovity always comes in convolution with chance oscillations in the natural process and there is an attractor, a fixed set of conditions in the future, that sets process direction.

9. Regarding the attractor's behaviour, we have a choice between two models:

1. Model I is the stationary Markov chain. Its stationary attractor is the climax state unique to the source community. In that state, transition

---

[12] I suggest to students, retreat to directed reading. Our paper with Madhur Anand and Kate He (1993) and my Statistical Ecology (2019, Chapter 20) contain much information including a full slate of references. Some downloadable gratis from my ResearchGate page https://www.researchgate.net/profile/Laszlo_Orloci2 . Latest versions can be ordered at cost from Amazon's Kindle Direct Publishing. Regarding choice of tutor, mathematical skills in Markov algebra help, but not enough. Tutor fluency in quantitative plant ecology is a must.

probability oscillations lose a sense of direction. Random events rule over compositional change. The longevity of the climax state is fixed by the longevity of the environmental conditions under which it arises.

2. Model II is the moving Markov chain. It is an option where the transition probabilities undergo a directional change in continuity within the time interval spanned by the analysis. In such a case the apropos objective seeks the functional form of change in transition probabilities in concordance with a change in the environmental predicates.

Stress-based statistical significance

As mentioned, the elements of **P**, Model I, are free from temporal directional change other than those arising by pure chance. This is not the same as saying that there is no directional change in **X** or **D**, and further, no directional change is free of random events occurring over the entire natural assembly process in the chronosere.

The test of statistical significance requires the observed $\sigma(D,\Delta)$ and the sampling distribution of the $\sigma(D,\Delta)$ function under the very un-natural assumption that the **X** and **D** are, in fact, free of temporal directional change. We can simulate this oxymoron by randomization (random sampling and resampling) of the elements in the observed **X** or by randomization of positions in the chronosere. Why do I not suggest deriving the sampling distribution from first principles that mathematical statisticians so compelled to do? My reason is simple. The distribution would have broad generality, but little local relevance (Orlóci 1980 2019). I opt for maximum local relevance for the practitioner; therefore, I elect the empiricism I find in Monte Carlo simulation.

Observational data (**X**) are input. Many recursive surges of randomization are involved and following each, a new $\sigma(D,\Delta)$ value is calculated. The elements of **D** are based on **X**. Thos of Δ are dased on the Markov scores fitted to the simulated **X**. I use the term "zero-order Markov" to characterize such a set of scores. The $\sigma(D,\Delta)$ distribution sought is constructed by ordering the simulated $\sigma(D,\Delta)$ values and counting percentages according to *% Simulated* $\sigma(D,\Delta) <= Observed$ $\sigma(D,\Delta)$. This distribution is specific to the no Markov-type directed change or zero *Markovity* hypothesis in the natural chronosere **X** of the source metacommunity.

To declare Markovity significant in the chronosere the value of the observed stress $\sigma(\mathbf{D}, \mathbf{\Delta})$ should be small. How small? For example, at the 5% significance level for Markovity we would consider the observed $\sigma(\mathbf{D}, \mathbf{\Delta})$ small if the simulated $\sigma(\mathbf{D}, \mathbf{\Delta})$ satisfied the condition *Simulated* $\sigma(\mathbf{D}, \mathbf{\Delta})$ <= *Observed* not more than 5% of the cases. In other words, the observed $\sigma(\mathbf{D}, \mathbf{\Delta})$ is considered *small* if it is a low probability event under the zero Markovity hypothesis. Setting critical values in significance testing is completely arbitrary. It should be remembered the decision becomes more conservative and have more rigour if the critical value is set to, say 1% or even smaller. In the other direction, the test becomes more liberal and loses rigour.

Since multiscaling is intended, a way to lay open more completely the total random effect, I perform the analysis at lag 1 to 20 in the example.[13] If I can show that the observed sharpness of Markovity exceeds what pure chance would allow with given probability *p*, Markovity is declared significant at the lag used. Verification of significance opens a wealth of information developed about stationary Markov chains that can be applied to the vegetation process. One of these is the convergence of the natural community process on a state we call the climax vegetation. Where Markovity is not found other models can be tried to probe for the mathematical function of the regularity suspected.

The complexities in the simulation are computational, needing a computer code which takes the chronosere **X** for input, performs Markov simulation, calculates stress values, and performs significance testing. I have written FITMARKO to do the calculations. The paleo relevés in **X** are arranged in the sequence set by the natural process: from bottom to tope in the sediment core.

---

[13] Remember 'lag' is the 'order' of Markovity. For example, relevés A, B, C, D, E are paired for distance calculation AB, AC, AD, AE, BC, ..., BE, CD, CE, or DE at lag one; AC, AE, BD at lag two; and AD, BE at lag three.

All intermediate results are stored on file when requested. The Markov part follows Orlóci et al. (1993, Orlóci 2019). Being conversational, FITMARKO's code expects users to have familiarity with my terminology. I suggest first use under tutoring.

Statistical significance testing

It is obvious from what I presented so far that there are different ways of doing the actual test, and not necessarily with identical outcome or precision. I summarise what FITMARKO does in sequence:

1. It creates a proxy of the zero-order Markov chain by randomization of the n paleorelevés in the chronosere or the palynomorphs among the positions. What is being done in both is equivalent to a process state where the palynomorph content of the paleorelevés is independent of time. In other words, the compositional change in the simulated community is rendered directionless by design.

2. A proxy chronosere created, the first simulated $\sigma(D, \Delta)$ is created. This is second I the set. The first is the observed $\sigma(D, \Delta)$.

3. It is tested whether the simulated $\sigma(D, \Delta)$ is less or equal to the observed $\sigma(D, \Delta)$. If that condition is true the number of stress values satisfying *simulated* $\sigma(D, \Delta)$ <= *observed* $\sigma(D, \Delta)$ is increased by one to 2.

4. Steps 1 to 3 are repeated a large number of times. Designate this number by N.

5. At this point there are N simulated $\sigma(D, \Delta)$ values at lag 1. Declare Markovity significant at lag 1 if the *number of simulated* $\sigma(D, \Delta)$ <= *observed* $\sigma(D, \Delta)$ value is small, say 5 or less per 100.

6. Repeat steps 1 to 5 until a pre-determined value of lag is reached.

To assist users with FITMARKO, I include a copy of the start-up dialogue in the analysis of the interglacial segment of Cwynar plant particle chronosere from 13038 BP to present. The input includes Cwynar's 11 dominant palynomorph taxa and 65 paleo relevés (Figure 2):

True BASIC Gold Edition — □

```
The MARKOV chain is fitted to an observed series. Input
data arranged by taxa or releves in a single column. Constant
time-step width is assumed. See the section 'post mortem' in the
printda file output by the program
Do you wish to proceed? -- press Y or N:y
y
C:\Users\lorlo\OneDrive\Documents\Tensor analysis\Hanging Lake\DATA 11 TAXA X 65 RELEVes BY TAXA.TRU
Should intermediate results be stored -- press Y or N:y
Data arranged by taxa?-press Y or N:y
Specify number of taxa: 11
Specify number of releves (length of the series): 65

   Specify LAG size upper limit 0,1,2, .:20
   Adjust data to equal releve totals - press Y or N:n
   Choose hypothesis to be tested:
   1- Ho: series is 0-order Markov (undirected); random permutation of positions used.
   2- Ho: series is m-order Markov; sampling/resampling of the stretched M used
   Choose 1 or 2:1
   Specify % error threshold:15
   Output file name extension: h111x651ag20err15
   Do data contain negative values - Y/N:n
   Is first record in the series where the Markov chain begins?- Y/N:n
   STEP SIZE: 1
=======================================================
Randomization begins. Specify number of iterations to be used:? 100|
```

In the question "Is first record in the series where the Markov chain begins? - Y/N: n" the word "first record" refers to the top sediment horizon. Response "n" indicates that the first record set in the data references the lowest sediment horizon sampled. After typing 100, an arbitrary number that appears, to me, reasonably large for this project, pressing the return key initiates the processing. The screen shows,

```
Iteration #: 100
STEP SIZE: 19
=========================================================

Iteration #: 100
STEP SIZE: 20
=========================================================

Iteration #: 100
Printda file:printdah111x651ag20err15.tru
Consult instruction at end of Printda file.
Press any key to exit.
```

Processing ends with the text appearing on the screen requesting "Press any key to exit."

There is access to the relevant PRINTDA file and others stored in the work folder. The file may take up hundreds of pages in a realistic case. Among these, some are useful to copy tables, locate errors, test long hand calculations in training use, implement revisions in the code, etc.

Example

The solution for *%number of simulated* $\sigma(D, \Delta) <= $ *observed* $\sigma(D, \Delta)$ at lag 1 to 20 is summarized in the table below --

Lag	Observed $\sigma(D,\Delta)$	%<=Observed $\sigma(D,\Delta)$	Lag	Observed $\sigma(D,\Delta)$	%<=Observed $\sigma(D,\Delta)$
1	0.9847	73	11	0.9139	40
2	0.9744	73	12	0.8839	44
3	0.9712	70	13	0.854	22

4	0.9493	76	14	0.8836	21
5	0.9499	67	15	0.8662	25
6	0.9385	66	16	0.864	21
7	0.9419	63	17	0.8328	18
8	0.9147	66	18	0.8624	17
9	0.9138	47	19	0.846	25
10	0.9075	54	20	0.8541	19

What do we read from this table? I explained it in general terms in the previous text. Now I explain it based on the text printed from the PRINTDA file of FITMARKO, rephrased with annotations:

The H_o tested states "The coenosere is undirected (zero-order Markov)". This means the observed value of 1- $\sigma(D, \Delta)$ has zero expectation. In other words, stress reigns, no trace of regularity other than random. The reference distribution **D** is based on the random-permuted **X** and Δ is based on the Markov scores from the randomised **X**. The probabilities, given as % values in the table, are coupled with probability points,

1%	5%	10%	25%	50%	75%	90%	95$	99%
.95956	.96951	.97144	.97899	.98291	.98475	.98528	.98551	.98573

The probability points are stress values. Reject H_o (zero-order Markovity) when the observed stress value is associates with a small probability.

Take the observed $\sigma(D,\Delta)$ = 0.9847 (lag 1). The corresponding per cent of simulated $\sigma(D,\Delta)$ values out of one hundred equal to or smaller than 0.9847 is 73. What can we say about this with certainty? The given chronosere has a significant level of stress. Yet, I will argue we cannot simply interpret "73" as an indication that the Interglacial section of the chronoseres' 11 selected taxa and 65 paleorelevés does not manifest Markov type directedness. I explain.

Consider the graphs in Figures 3 and 4. The graphs show the observed and generated $\sigma(D,\Delta)$ values one per each step of lag from 1 to 20. The reader is reminded, our basic data set in this part of the exercise includes 11 taxa in 65 relevés. Remember, the period length covers the Interglacial from year 13068 BP to present.

What can we read from Figures 3 about process Markovity during the Interglacial period? For one thing, the large stress values tell us of a Markovity weak but gain strength with increasing lag. What can we do with this bit of information? If we examine the regression statistics in Figure 3 and Appendix 3a, we find parameter b=-0.00767008:

Parm	Value	Std Error	t-value	90% Confidence Limits		P>|t|
a	0.985875789	0.005932724	166.1759134	0.975588069	0.99616351	0.00000
b -	0.00767008	0.000495254	-15.4871479	-0.00852888	-0.00681127	0.00000

Coefficient *b* is a tangent value. It indicates a modest but statistically highly significant 0.4 degrees slope, *i.e.* P>|t| ≈ 0.00000. Read this: "the probability of a *t* value that chance is handing us when in fact the true *b* is zero". The individual stress values are large, consequently, we regard Markovity weak but not insignificant. These are telling us simply, we have in hands a process that is Markov whose statistical significance rises in strength with increasing lag. The probabilistic aspect is emphasized in Figure 4.

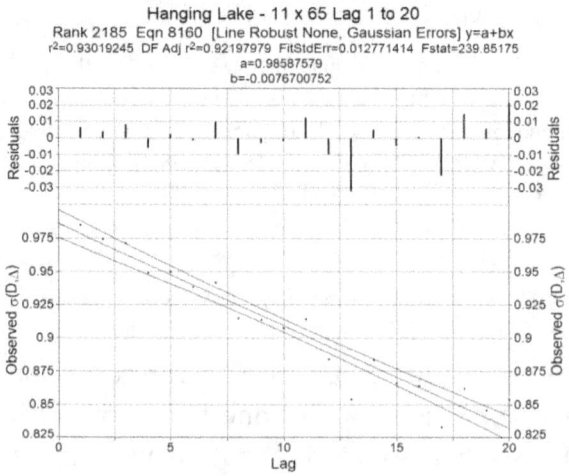

Figure 3. Relationship of lag and observed σ(**D,Δ**). Each point represents a state of lag at which the analysis of the 65 relevé by 11 taxa data set yielded the σ(**D,Δ**) value. The regression line, residuals, and 90% confidence limits are shown. The numerical results in the caption and the Appendices 3a,b reveal highly significant regression statistics.

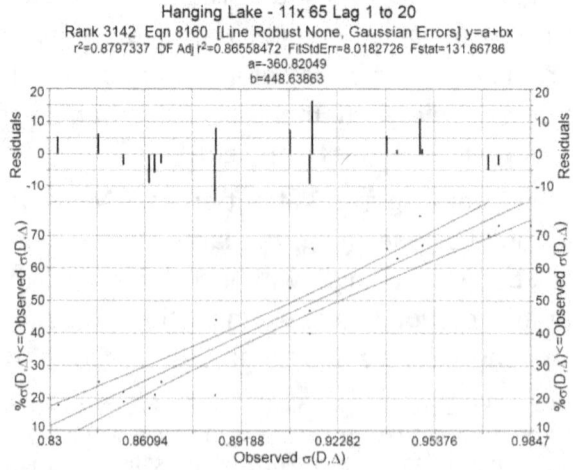

Figure 4. Relationship of observed σ(**D,Δ**) and the % number of generated σ(**D,Δ**) less or equal to the observed. Points represent different cases of lag. The regression line, residuals, and 90% confidence limits are shown. The numerical results in caption and Appendix 4a,b reveal highly significant regression statistics.

The same analysis is performed on the 27 dominant taxa by 65 relevés data set and found similar Markovian regularity. This is how I see the relevance of these to syndynamics:

1. The high sigma values indicate dominant random effects. When this associates with significant Markovity we see verified the expected: a Markov attractor in chaos. The vast regional extent of the plant particles' source and natural particle dispersal intensify the random effect. It intensifies the fuzziness of the taxonomic scheme which entombs even deeper in randomness the directed regularity.

2. The outcome shows, it makes sense to examine Markovity as a multiscale phenomenon. lag is the scale variable at present. In the more restricted sense, lag is Markovity's order. There are other methods of multiscaling. Anyone who thinks about it will certainly come up with a new one. I suggest a nested-hierarchical frame (Orlóci 2009, 2012, 2019 Section 12.5 and Chapter 19).

3. We should take the decline of stress with increased lag as reality and consider Markovity a recognized mechanical component of the long-term vegetation process. Significant as a component trend, but in stress terms rather weak. We can say, the Markov attractor oscillates randomly without directed regularity. In such a state the attractor state is syndynamic's climax. Perturbation is needed for the process to be jolted out of the climax state. Global climate warming is doing just that to the Beringian vegetation.

4. The effect of ongoing global climate warming is visible in the entire arctic zone - Arid Steppe, Tundra, Taiga and Boreal forest. This may have left a detectable trace in the plant particle spectrum being created in the Hanging Lake sediment since Cwynar completed his sampling almost 4 decades ago. Would it be in too short of an elapsed time for that? We can look at this more closely. We take Professor Cwynar's (1982b) carbon dates and sediment depths and create the two graphs in Figure 5.

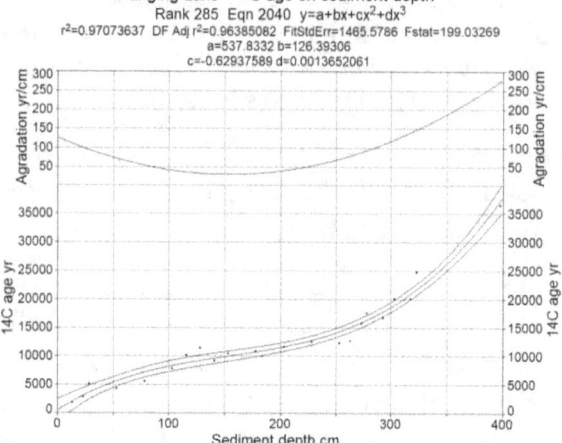

Figure 5. Sediment aggradation by Cwynar's measurements. The graphs are mine. Curves: aggradation (1st differential equation yr/cm units), regression line, 90% confidence limits. See the complete set of numerics in Appendices 5a and b and explanations in the sequel. How to read this graph? It links time and sediment depth with the rate of change.

Inspection of the graph in Figure 5 is revealing. The 1st derivative (Appendices 5a) expresses the aggradation rate in the manner of how many years it takes for the sediment thickness to increase by 1 cm:

1st Deriv min	X-Value	1st Deriv max	X-Value
29.676525425	153.67054829	274.17215699	398

The minimum rate required in round figures 30 years at sediment depth 154 cm 10500 years ago. This rate lasted from about 11000 to 9000 BP. The graph indicates also a substantial reduction in the number of years per 1cm aggradation under warming cycles and substantial increase under cooling cycles. The present rate is 125 years. This implies about 3.04 mm rise of the sediment depth in the lake since 1982. 5. The residuals in Figures 4 and 5 deserve attention. Regression analysis isolates the trended variation captured by the regression line from all others in the grab bag called random variation. The unexplained or unexplainable variation could be better terms. The amount is about 12 % in Figure 4 and 7 % in Figure 5. The numerics come from the analysis of variance table in Appendix 4a and Appendix 5a.

Climax prediction

What to predict? – for us, the compositions of the attractor. Since the process that points to the attractor is defined by the transition probability matrix, we iterate $M_{i+1}=M_iP$ recursively to converge on a point M, my symbol for the attractor. We recognise that point being approached when the difference of the generated Markov relevés M_i-M_{i+1} flattens, or equivalently the first derivative of the function written for the difference M_i-M_{i+1} approaches the zero line. This is seen in Figures 6a and 6b. The numerical results are in Appendices 6aa and ab, 6ba and bb, and Tables 5 and 6.

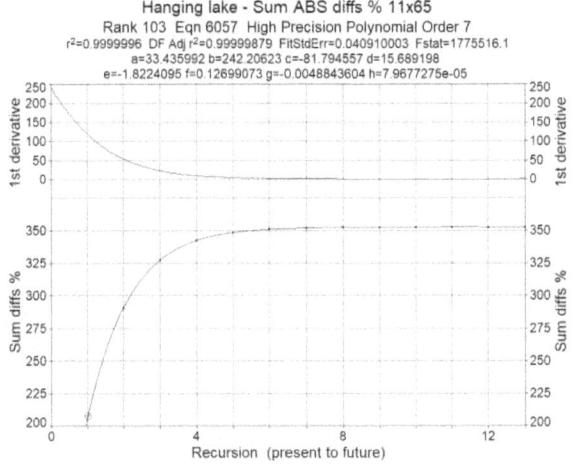

Figure 6a. Analysis of the sum of absolute (ABS) per cent differences (last line in Table 5). The top curve is the map of the 1st differential equation. The bottom curve is the regression line fitted to the sum per cent differences. It flattens at about M_5 where the derivative approaches the zero line. Take this as the predicted entry point of the regional metacommunity into the climax state (period), about 1200 years in the future.

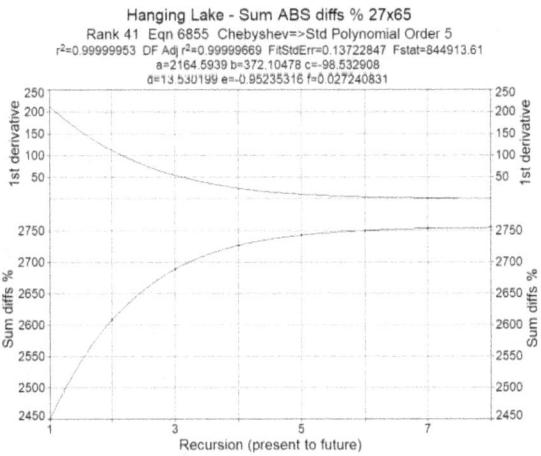

Figure 6b. The graph has the same design as Figure 6a, except the analysis involves 27 characteristic palynomorph taxa and 65 paleorelevés. The last line of Table 6 is portrayed.

The transition probability matrix **P** is defined in both cases for the Interglacial portion of the Cwynar chronosere including 11 palynomorph taxa (Figure 1) and 65 paleo relevés for the case in Figure 6a and 27 palynomorph taxa in 65 paleorelevés for the case in Figure 6b.

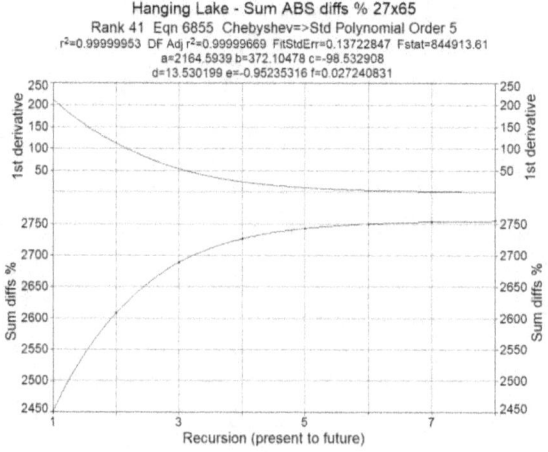

Hanging Lake - Sum ABS diffs % 27x65
Rank 41 Eqn 6855 Chebyshev=>Std Polynomial Order 5
r²=0.99999953 DF Adj r²=0.99999669 FitStdErr=0.13722847 Fstat=844913.61
a=2164.5939 b=372.10478 c=-98.532908
d=13.530199 e=-0.95235316 f=0.027240831

Figure 6b. The graph has the same design as Figure 6a, except the analysis involves 27 characteristic palynomorph taxa and 65 paleorelevés. The last line of Table 6 is portrayed.

Table 5. Transition probability matrix **P** and generated Markov relevés **M**, based on 11 palynomorph taxa in 65 paleo relevés spanning the last 13068 years. The case is portrayed in Figure 6a. The **P** matrix is an average of 64 transition probability matrices at lag 1 paired in chronological order. The rows of **P** have unit sums. M_o is the paleorelevé X_1 of the top sediment layer ("present" 1982). Relevé X_{65} corresponds to the sediment layer at 244 cm depth Abbreviations for taxa are Cwynar's.

(a) Transition probability matrix **P**

	ID	1	2	3	4	5
1	2 Aln.v	0.45354	0.34464	0.01216	0.03275	0.0531
2	5 Bet	0.1353	0.61238	0.01301	0.03924	0.045
3	8 Ercae	0.126	0.35026	0.36712	0.02812	0.03623
4	9 Erles	0.12251	0.36593	0.00981	0.37111	0.03568
5	12 Pic	0.1505	0.34483	0.00996	0.02796	0.36923
6	15 Slx	0.06327	0.40698	0.0071	0.0211	0.02177
7	23 Art	0.07804	0.35314	0.00684	0.01906	0.02451

8	27 Braae	0.11924	0.34152	0.00854	0.0222	0.03314
9	30 Ch/Am	0.06032	0.36055	0.00524	0.01491	0.02219
10	32 Cypae	0.10927	0.38418	0.00902	0.02676	0.03483
11	54 Poaae	0.10828	0.35108	0.00839	0.02464	0.03189

	ID	6	7	8	9	11	11
1	2 Aln.v	0.0149	0.00783	0.00046	0.00025	0.05215	0.02822
2	5 Bet	0.03605	0.01148	0.00053	0.00047	0.07191	0.03463
3	8 Ercae	0.01759	0.00654	0.00035	0.00022	0.04472	0.02284
4	9 Erles	0.01793	0.00646	0.00034	0.00022	0.04652	0.02349
5	12 Pic	0.01556	0.00668	0.00042	0.00025	0.04955	0.02506
6	15 Slx	0.40322	0.00967	0.00031	0.00045	0.04346	0.02267
7	23 Art	0.02707	0.42336	0.00043	0.00054	0.04191	0.02507
8	27 Braae	0.02273	0.01242	0.35913	0.00053	0.04821	0.03233
9	30 Ch/Am	0.0333	0.01482	0.00053	0.41794	0.04198	0.02823
10	32 Cypae	0.02094	0.00745	0.0004	0.00031	0.383	0.02385
11	54 Poaae	0.02144	0.00873	0.00041	0.0004	0.04499	0.39974

(b) Markov relevés (estimates of attractor composition) M_1 to M_8 at lag 1 generated recursively by $M_{i+1} = M_i P$. Note, $M_1 = M_0 P$.

Lag 1 11x65

ID	Mo	M1	M2	M3	M4
2 Aln.v	559	390.466	335.320	316.717	310.266
5 Bet	532	717.130	767.600	781.897	786.136
8 Ercae	44	34.427	31.153	29.979	29.543
9 Erles	78	80.315	82.279	83.173	83.509
12 Pic	149	121.642	111.070	106.893	105.227
15 Slx	33	50.613	61.312	66.557	68.914
23 Art	32	27.792	26.803	26.614	26.604
27 Braae	1	1.111	1.163	1.184	1.191
30 Ch/Am	1	0.974	1.008	1.036	1.052
32 Cypae	148	142.636	144.468	145.993	146.736
54 Poaae	60	69.894	74.823	76.956	77.820

ID	Mo	M5	M6	M7	M8
2 Aln.v	559	307.974	307.143	306.836	306.721
5 Bet	532	787.457	787.890	788.039	788.091

8 Ercae	44	29.377	29.313	29.288	29.278
9 Erles	78	83.620	83.654	83.663	83.664
12 Pic	149	104.562	104.296	104.190	104.148
15 Slx	33	69.929	70.354	70.530	70.602
23 Art	32	26.622	26.638	26.647	26.652
27 Braae	1	1.194	1.195	1.195	1.195
30 Ch/Am	1	1.061	1.065	1.067	1.067
32 Cypae	148	147.044	147.163	147.207	147.223
54 Poaae	60	78.158	78.287	78.336	78.355

(c) Deviations M_i-M_o, i=1 to 8

Lag 1 11x65

ID	Mo	M1-Mo	M2-Mo	M3-Mo	M4-Mo
2 Aln.v	559	-168.534	-223.680	-242.283	-248.734
5 Bet	532	185.130	-235.600	-249.897	-254.136
8 Ercae	44	-9.573	-12.847	-14.021	-14.457
9 Erles	78	2.315	4.279	5.173	5.509
12 Pic	149	-27.358	-37.930	-42.107	-43.773
15 Slx	33	17.613	28.312	33.557	35.914
23 Art	32	-4.208	-5.197	-5.386	-5.396
27 Braae	1	0.111	0.163	0.184	0.191
30 Ch/Am	1	-0.026	0.008	0.036	0.052
32 Cypae	148	-5.364	-3.532	-2.007	-1.264
54 Poaae	60	9.894	14.823	16.956	17.820

ID	Mo	M5-Mo	M6-Mo	M7-Mo	M8-Mo
2 Aln.v	559	-251.026	-251.857	-252.164	-252.279
5 Bet	532	-255.457	-255.890	-256.039	-256.091
8 Ercae	44	-14.623	-14.687	-14.712	-14.722
9 Erles	78	5.620	5.654	5.663	5.664
12 Pic	149	-44.438	-44.704	-44.810	-44.852
15 Slx	33	36.929	37.354	37.530	37.602
23 Art	32	-5.378	-5.362	-5.353	-5.348
27 Braae	1	0.194	0.195	0.195	0.195
30 Ch/Am	1	0.061	0.065	0.067	0.067
32 Cypae	148	-0.956	-0.837	-0.793	-0.777
54 Poaae	60	18.158	18.287	18.336	18.355

(d) ABS(100(M_i-M_o)/ M_o) %, i=1 to 8
Lag 1 11x65

ID	Mo	M1-Mo %	M2-Mo %	M3-Mo %	M4-Mo %
2 Aln.v	559	30	40	43	44
5 Bet	532	35	44	47	48
8 Ercae	44	22	29	32	33
9 Erles	78	3	5	7	7
12 Pic	149	18	25	28	29
15 Slx	33	53	86	102	109
23 Art	32	13	16	17	17
27 Braae	1	11	16	18	19
30 Ch/Am	1	3	1	4	5
32 Cypae	148	4	2	1	1
54 Poaae	60	16	25	28	30
Sum	1637	208	291	327	342

ID	Mo	M5-Mo %	M6-Mo %	M7-Mo %	M8-Mo %
2 Aln.v	559	45	45	45	45
5 Bet	532	48	48	48	48
8 Ercae	44	33	33	33	33
9 Erles	78	7	7	7	7
12 Pic	149	30	30	30	30
15 Slx	33	112	113	114	114
23 Art	32	17	17	17	17
27 Braae	1	19	20	20	20
30 Ch/Am	1	6	6	7	7
32 Cypae	148	1	1	1	1
54 Poaae	60	30	30	31	31
Sum	1637	348	351	352	352

Each Markov relevé is by proxy a valid estimate of the source meta-community's future composition as the process is approaching the regional climax. Figure 6 shows the generated Markov series up to 8 steps in the future. Since we are interested to predict the regional climax, we must select an **M** for the estimate. I choose M_6, the point where we already see advanced convergence. This is where the "sum ABS diff %" function is levelling off, or equivalently, the 1st derivative drops close to zero and keeps tracking that trend. Many adverse events may change the climax condition over the period involved, an estimated 1200 years. These would change the expected **P** that would make invalidity the estimate M_6. I answer two questions:

1. Will the transition probability matrix hold valid for 1200 years? This is very unlikely considering the ongoing climate warming process.

2. Why 200 years for step width? The value that comes from $13068/64 \approx 200$, the average step length for which **P** is defined.

The same calculations performed on 27 selected taxa of 65 paleorelevés gave us the results in Table 6 and Figure 6b.

Table 6. Same design as Table 5, except, the analysis involves 27 characteristic palynomorph taxa in 65 paleorelevé. Transition probability matrix is not given. The sum of absolute differences, the base of Figure 6b, are in the last line of the table. ID number and code copied from original records.

(b) Markov relevés M_1 to M_8 at lag 1 generated recursively by $M_{i+1} = M_i P$
Lag 1 27x65

ID	Mo=Xo	M1	M2	M3	M4
5 Bet	532	707.184	756.377	770.844	775.335
15 Slx	33	48.279	57.365	61.888	63.979
10 Led	10	18.842	22.613	24.126	24.712
54 Paae	60	67.735	71.359	72.860	73.442
74 Equ	7	11.338	14.710	16.622	17.592
18 Vac	12	16.641	18.325	18.897	19.079
81 Sel.si	1	4.321	5.278	5.536	5.597
1 Aln.i	2	4.324	5.213	5.530	5.639
77 Lyc.a	0	1.751	2.428	2.677	2.767
83 Spg	51	50.846	52.022	52.764	53.115
65 Sax.hi	0	1.287	1.689	1.811	1.847
57 Ptl	0	0.942	1.398	1.598	1.682
59 Ranae	3	4.133	4.475	4.570	4.592
9 Erls	78	78.295	79.079	79.444	79.562
29 Cryae	0	0.592	0.822	0.908	0.941

52 Pln.ca	0	0.184	0.340	0.454	0.532
80 Plpae	3	2.994	3.227	3.382	3.460
27 Braae	1	1.086	1.125	1.139	1.144
6 Csp	6	6.077	6.107	6.109	6.104
30 Ch/Am	1	0.946	0.957	0.973	0.983
24 Astae	2	1.697	1.607	1.579	1.570
26 Asg	2	1.111	0.775	0.647	0.598
23 Art	32	27.092	25.524	25.009	24.838
32 Cypae	148	138.690	138.205	138.705	139.039
8 Ercae	44	33.719	29.985	28.583	28.044
12 Pic	149	118.750	106.829	102.033	100.082
2 ALN	559	387.158	328.191	307.346	299.769

ag 1 27x65

ID	Mo=Xo	M5	M6	M7	M8
5 Bet	532	776.811	777.324	777.513	777.586
15 Slx	33	64.910	65.316	65.492	65.567
10 Led	10	24.934	25.017	25.048	25.060
54 Paae	60	73.658	73.736	73.763	73.772
74 Equ	7	18.058	18.275	18.375	18.420
18 Vac	12	19.133	19.147	19.150	19.149
81 Sel.si	1	5.608	5.608	5.606	5.605
1 Aln.i	2	5.675	5.687	5.691	5.692
77 Lyc.a	0	2.798	2.810	2.814	2.815
83 Spg	51	53.264	53.323	53.346	53.355
65 Sax.hi	0	1.858	1.861	1.861	1.861
57 Ptl	0	1.717	1.731	1.736	1.739
59 Ranae	3	4.596	4.595	4.594	4.593
9 Erls	78	79.586	79.585	79.579	79.575
29 Cryae	0	0.952	0.957	0.958	0.959
52 Pln.ca	0	0.584	0.618	0.640	0.655
80 Plpae	3	3.496	3.511	3.517	3.520
27 Braae	1	1.145	1.146	1.146	1.146
6 Csp	6	6.099	6.096	6.094	6.093
30 Ch/Am	1	0.989	0.991	0.993	0.994
24 Astae	2	1.567	1.566	1.565	1.565
26 Asg	2	0.579	0.572	0.570	0.569

23 Art	32	24.781	24.763	24.758	24.757
32 Cypae	148	139.192	139.253	139.275	139.283
8 Ercae	44	27.833	27.749	27.716	27.702
12 Pic	149	99.284	98.957	98.822	98.766
2 ALN	559	296.944	295.866	295.445	295.278

(c) Deviations M_i-M_o

Lag **1 27x6**

ID	Mo=Xo	M1-Mo	M2-Mo	M3-Mo	M4-Mo
5 Bet	532	175.184	224.377	238.844	243.335
15 Slx	33	15.279	24.365	28.888	30.979
10 Led	10	8.842	12.613	14.126	14.712
54 Paae	60	7.735	11.359	12.860	13.442
74 Equ	7	4.338	7.710	9.622	10.592
18 Vac	12	4.641	6.325	6.897	7.079
81 Sel.si	1	3.321	4.278	4.536	4.597
1 Aln.i	2	2.324	3.213	3.530	3.639
77 Lyc.a	0	1.751	2.428	2.677	2.767
83 Spg	51	-0.154	1.022	1.764	2.115
65 Sax.hi	0	1.287	1.689	1.811	1.847
57 Ptl	0	0.942	1.398	1.598	1.682
59 Ranae	3	1.133	1.475	1.570	1.592
9 Erls	78	0.295	1.079	1.444	1.562
29 Cryae	0	0.592	0.822	0.908	0.941
52 Pln.ca	0	0.184	0.340	0.454	0.532
80 Plpae	3	-0.006	0.227	0.382	0.460
27 Braae	1	0.086	0.125	0.139	0.144
6 Csp	6	0.077	0.107	0.109	0.104
30 Ch/Am	1	-0.054	-0.043	-0.027	-0.017
24 Astae	2	-0.303	-0.393	-0.421	-0.430

ID	Mo=Xo			
26 Asg	2	-0.889	-1.225	-1.353	-1.402
23 Art	32	-4.908	-6.476	-6.991	-7.162
32 Cypae	148	-9.310	-9.795	-9.295	-8.961
8 Ercae	44	-10.281	-14.015	-15.417	-15.956
12 Pic	149	-30.250	-42.171	-46.967	-48.918
2 ALN	559	-171.842	-230.809	-251.654	-259.231

Lag **1 27x65**

ID	Mo=Xo	M5-Mo	M6-Mo	M7-Mo	M8-Mo
5 Bet	532	244.811	245.324	245.513	245.586
15 Slx	33	31.910	32.316	32.492	32.567
10 Led	10	14.934	15.017	15.048	15.060
54 Paae	60	13.658	13.736	13.763	13.772
74 Equ	7	11.058	11.275	11.375	11.420
18 Vac	12	7.133	7.147	7.150	7.149
81 Sel.si	1	4.608	4.608	4.606	4.605
1 Aln.i	2	3.675	3.687	3.691	3.692
77 Lyc.a	0	2.798	2.810	2.814	2.815
83 Spg	51	2.264	2.323	2.346	2.355
65 Sax.hi	0	1.858	1.861	1.861	1.861
57 Ptl	0	1.717	1.731	1.736	1.739
59 Ranae	3	1.596	1.595	1.594	1.593
9 Erls	78	1.586	1.585	1.579	1.575
29 Cryae	0	0.952	0.957	0.958	0.959
52 Pln.ca	0	0.584	0.618	0.640	0.655
80 Plpae	3	0.496	0.511	0.517	0.520
27 Braae	1	0.145	0.146	0.146	0.146
6 Csp	6	0.099	0.096	0.094	0.093
30 Ch/Am	1	-0.011	-0.009	-0.007	-0.006
24 Astae	2	-0.433	-0.434	-0.435	-0.435
26 Asg	2	-1.421	-1.428	-1.430	-1.431
23 Art	32	-7.219	-7.237	-7.242	-7.243
32 Cypae	148	-8.808	-8.747	-8.725	-8.717
8 Ercae	44	-16.167	-16.251	-16.284	-16.298
12 Pic	149	-49.716	-50.043	-50.178	-50.234

| 2 ALN | 559 | -262.056 | -263.134 | -263.555 | -263.722 |

(d) ABS(100(M_i-M_o)/ M_o) %, i=1 to 8

Lag **1 27x65**

ID	Mo=Xo	M1-Mo %	M2-Mo %	M3-Mo %	M4-Mo %
5 Bet	532	33	42	45	46
15 Slx	33	46	74	88	94
10 Led	10	88	126	141	147
54 Paae	60	13	19	21	22
74 Equ	7	62	110	137	151
18 Vac	12	39	53	57	59
81 Sel.si	1	332	428	454	460
1 Aln.i	2	116	161	176	182
77 Lyc.a	0	0	0	0	0
83 Spg	51	0	2	3	4
65 Sax.hi	0	0	0	0	0
57 Ptl	0	0	0	0	0
59 Ranae	3	38	49	52	53
9 Erls	78	0	1	2	2
29 Cryae	0	0	0	0	0
52 Pln.ca	0	0	0	0	0
80 Plpae	3	0	8	13	15
27 Braae	1	9	12	14	14
6 Csp	6	1	2	2	2
30 Ch/Am	1	5	4	3	2
24 Astae	2	15	20	21	22
26 Asg	2	44	61	68	70
23 Art	32	15	20	22	22
32 Cypae	148	6	7	6	6
8 Ercae	44	23	32	35	36
12 Pic	149	20	28	32	33
2 ALN	559	31	41	45	46
Sum	1736	939	1300	1437	1489

Lag **1 27x65**

ID	Mo=Xo	M5-Mo %	M6-Mo %	M7-Mo %	M8-Mo %
5 Bet	532	46	46	68	46

15 Slx	33	97	98	51	99
10 Led	10	149	150	40	151
54 Paae	60	23	23	81	23
74 Equ	7	158	161	38	163
18 Vac	12	59	60	63	60
81 Sel.si	1	461	461	18	461
1 Aln.i	2	184	184	35	185
77 Lyc.a	0	0	0	0	0
83 Spg	51	4	5	96	5
65 Sax.hi	0	0	0	0	0
57 Ptl	0	0	0	0	0
59 Ranae	3	53	53	65	53
9 Erls	78	2	2	98	2
29 Cryae	0	0	0	0	0
52 Pln.ca	0	0	0	0	0
80 Plpae	3	17	17	85	17
27 Braae	1	15	15	87	15
6 Csp	6	2	2	98	2
30 Ch/Am	1	1	1	101	1
24 Astae	2	22	22	128	22
26 Asg	2	71	71	349	72
23 Art	32	23	23	129	23
32 Cypae	148	6	6	106	6
8 Ercae	44	37	37	159	37
12 Pic	149	33	34	151	34
2 ALN	559	47	47	189	47
Sum	1736	1509	1516	2236	1520

I summarize the prediction section. Starting with Cwynar's 1982 plant particle counts in the Hanging Lake sediment core's top horizon X_0, 8 predictions were generated for the attractor in 200-year steps from 1982 into the future up to 3582 AD. The time scale of the Markov prediction, just as much as the time scale for the ongoing global climate warming cycle, is considerably below the Deka millennial time scale on which a considerably more consequential process is unfolding. It is the inevitable return of a new Ice Age in the Northern Hemisphere for which the principle forcing mechanism is celestial mechanics, clearly explained by Milankovitch (1947). No anthropogenic effect that appears to be the main reason behind the ongoing global climate warming could stop the return to a new Ice Age in due time to the large landmass of the Northern hemisphere. Should the immediate future comes as predicted based on Markov considerations, what kind of change should we expect in the source vegetation of Hanging Lake's particle collection? We can read that from the columns of Table 7.

Table 7. Summary of Markov predictions at lag 1 for attractor state M_8 (3582 AD) relative to X_o. Columns 4, 5 are taken from Table 5, and columns 6, 7 from Table 6.

Lag 1			11X65		27x65		
ID CODE	Mo=Xo	Change	%	Change	%	Site type
5 Bet	532	256.091	-48	245.586	-46	Tundra, Cold Steppe
15 Slx	33	37.602	114	32.567	99	Wetland
10 Led	10	--	--	15.060	151	Wetland
54 Poaae	60	18.355	31	13.772	23	Wetland
74 Equ	7	--	--	11.420	163	Wetland
18 Vac	12	--	--	7.149	60	Taiga, Boreal Forest
81 Sel.si	1	--	--	4.605	461	Rocky Tundra
1 Aln.i	2	--	--	3.692	185	Wetland
77 Lyc.a	0	--	--	2.815	--	Taiga, Boreal Forest
83 Spg	51	--	--	2.355	5	Wetland
65 Sax.hi	0	--	--	1.861	--	Cold Steppe
57 Ptl	0	--	--	1.739	--	Cold Steppe
59 Ranae	3	--	--	1.593	53	Wetland
9 Erles	78	5.664	7	1.575	2	Taiga, Boreal Forest
29 Cryae	0	--	--	0.959	--	Cold Steppe
52 Pln.ca	0	--	--	0.655	--	Cold Steppe
80 Plpae	3	--	--	0.520	17	Cold Steppe
27 Braae	1	0.195	20	0.146	15	Cold Steppe
6 Csp	6	--	--	0.093	2	Tundra
30 Ch/Am	1	0.067	7	-0.006	-1	Cold Steppe
24 Asta	2	--	--	-0.435	-22	Cold Steppe
26 Asg	2	--	--	-1.431	-72	Cold Steppe
23 Art	32	-5.348	-17	-7.243	-23	Cold Steppe
32 Cypae	148	-0.777	-1	-8.717	-6	Wetland
8 Ercae	44	-14.722	-33	-16.298	-37	Taiga, Boreal Forest

12 Pic	149	-44.852	-30	-50.234	-34	Taiga, Boreal Forest
2 Aln.v	559	-252.279	-45	-263.722	-47	Alpine, Sub-boreal

Regarding the performance of the taxa, Table 7 identifies losers, neutrals, and gainers. The table suggests that postglacial climate warming has conditioned the transition probability matrix and by that way defined the tone of the prediction. Under such conditions, the taxa characteristic for the rocky Tundra, wetlands, and the mesic segment of the environmental gradient are increasing their representation.

Comparison of Vostok and Renland

Just as a reminder, the Vostok and Renland records are public domain. The temperature values are measured indirectly based on deuterium (heavy Hydrogen, D) extracted from air bubbles in the ice. Since the air bubbles are transported by precipitation from the inversion layer where the water drops are formed, the measured temperature is site inversion-layer specific. A consequence of this is that the temperature is replaced by the difference of the historic value and the current D-based average temperature at the site.

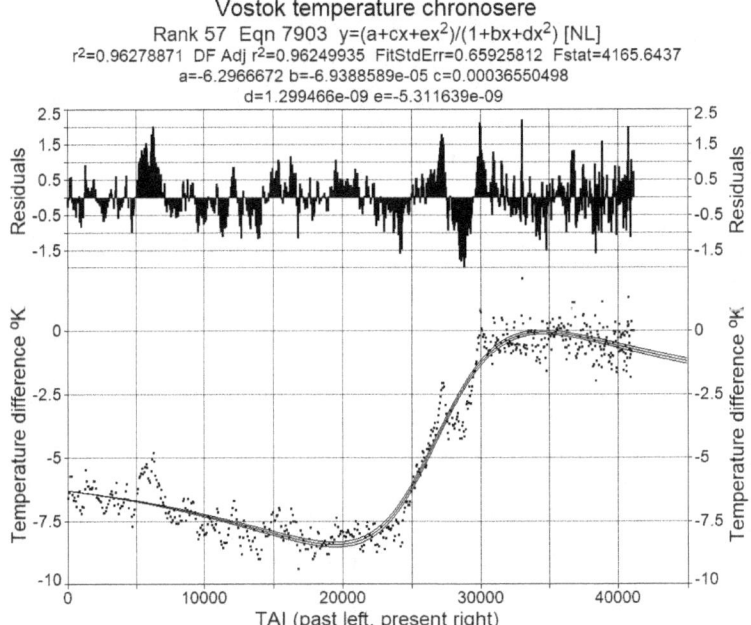

Vostok temperature chronosere
Rank 57 Eqn 7903 $y=(a+cx+ex^2)/(1+bx+dx^2)$ [NL]
$r^2=0.96278871$ DF Adj $r^2=0.96249935$ FitStdErr=0.65925812 Fstat=4165.6437
a=-6.2966672 b=-6.9388589e-05 c=0.00036550498
d=1.299466e-09 e=-5.311639e-09

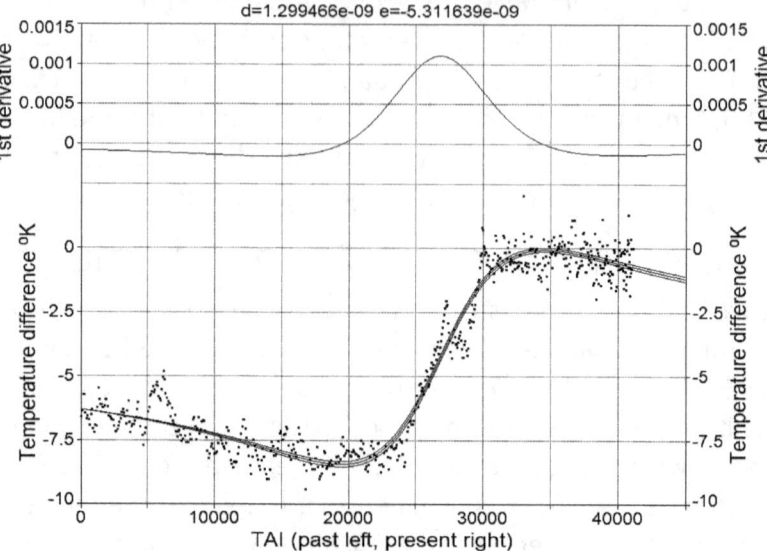

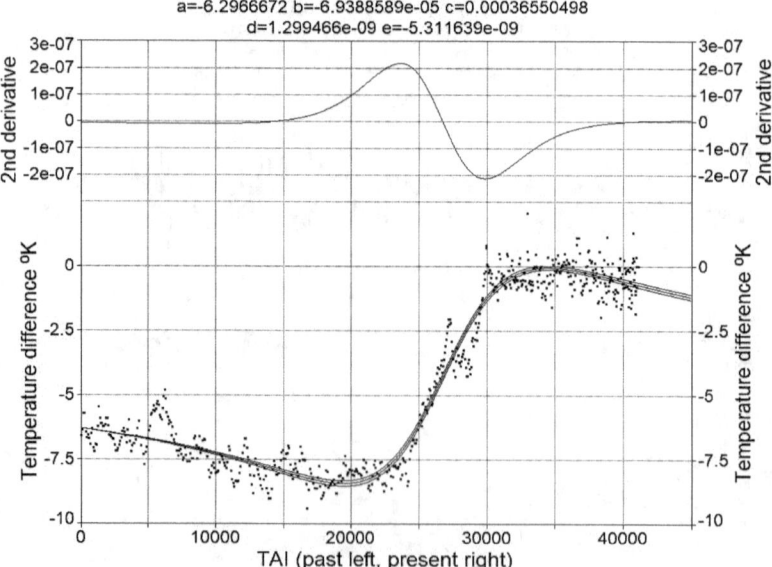

Figure 7. A segment of the Petit's Vostok temperature chronosere covering 41112 ice core years. The time of the period's initiation (TAI) is the zero point, present (1982) is at 41112 at the TAI scale. Temperatures are deuterium based expressed as

deviations. See the main text for technical details. The graphs' base advance regression analysis by the application TableCurve. See the numeric tables in Appendix 7.

The Vostok chronosere is presented in the regression graph of Figure 7. The figure has four basic elements:

1. Regression line. The equations and regression statistics are found in the figure's caption and more in the tables of numerical summary in Appendix 7.

2. Graph of residuals. These are distances measured from the regression line vertically to the sample points; in all, 629 points are shown. The deviations are symmetrically distributed around the regression line and portray temperature oscillations in time. I refer to the residuals' as random oscillations until otherwise determined. This emphasizes on effects beyond the disciplined regularity imposed by time on a chaotic system. The Vostok temperature oscillations have considerable amplitude along the entire period length of the chronosere.

3. Graph of 1st differential equation.

4. Graph of 2nd differential equation.

I should not pass this opportunity to mention that ecologists are used to seeing others use calculus starting with the idea that there is change and write it up as a rate or a differential equation. A few will take the exercise one big step further and use integral calculus to find the source equation. If they did, all would see the anaemic nature of the effort trying to reveal complex behaviour or relationship from elementary manifestations. What I do here is very different. I take information-rich data and first build an integral function. I use differential calculus to learn more about the source equation, such as finding extreme (characteristic) points. This is completely consistent with my commitment to a holistic approach in this exercise. The mathematical difficulty is minimised by available application programs of high quality.

Considering this further, my experimentally defined integral equation is

$$f(x)=\frac{a+cx+ex^2}{1+bx+dx^2} \quad (9)$$

Differentiation[14] gives me the 1st and 2nd differential equations,

---

[14] David Scherfgen's Derivative Calculator http://www. derivative-calculator.net/

$$\frac{d}{dx}f(x)=\frac{-(cdx^2-ebx^2+2adx-2cx-c+ab)}{(dx^2+bx+1)^2} \quad (10)$$

$$\frac{d^2}{d^2x}f(x)=\frac{2(cd^2x^3-ebdx^3+3ad^2x^2-3edx^2-3cdx+3abdx-ad-bc+ab^2+e)}{(dx^2+bx+1)^3} \quad (11)$$

If readers disagree, please let me know their point (lorloci@uwo.ca).

Some observations are in order, to reflect on the historic warming rate during the period that leads out of the deepest Ice Age to the beginning of the Interglacial following the melt of the Wisconsin ice sheet. I use the extreme points of the 2nd differential equation as my markers of the period start and end points within which I do the arithmetic:

2nd deriv max rear 23661 TAI (time after the period's initiation)
Observed °K -8.27
2nd deriv min year 29901 TAI
Observed °K -0.18
I divide this period into two based on the midterm anomaly (see temperature graph):
From year 23661 TAI to year 27310 TAI
Period length 3649 years
Observed temperature yr 27310 °K -2.04
Temperature difference °K 8.27-2.04=6.23 or 1.71 °K per 100 years
Year 29901 TAI to year r 27660 TAI
Period length 2241 years
Observed temperature year 27660 °K-4.28
Temperature difference °K 4.28-0.18=4.10 or 1.83 °K per 100 years

The warming rates are minuscule compared to the rates attributed to the ongoing climate warming. Yet small rates that are persisting over an aggregated 6k years had the power to melt the continental ice shield thousands of meters down, and allow a period of reconstruction began in the zonal pattern of the biota on the emerging terra novae over vast areas under full ecological co-ordination.

Yet another point I should make is regarding Schweingruber's (1996) book "Dendroecology". The book includes a graph under Schwinegruber's Figure 6 (my Figure 8). This helps me to make visual the global deuterium gradient in temperature terms superimposing locations from the Yukon Territory to Puerto Rico. Based on such a graph the zonal pattern of the biota can be reconstructed along the equator to the arctic latitudinal gradient.

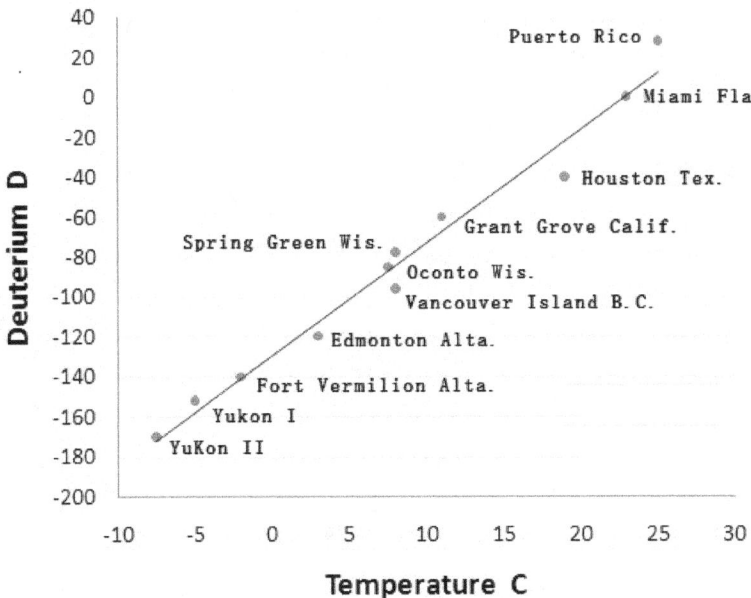

Figure 8. Temperature to deuterium conversion adopted from Schweingruber (1996). Deuterium is taken from the cellulose of tree rings.

The simple linear relationship of deuterium and ambient temperature is striking. Schweingruber's original graph and text gives guidance regarding entities and method of conversion. Using the regression equation (Figure 8) D=5.666T -129.3, for which R^2=0.976, local surface temperature can be estimated on the line by $t = \dfrac{D + 129.3}{5.666}$, the same in both °K or °C units. The D value is standardized. Relevant to matters concerning the use of deuterium as a temperature proxy, I mention works by Grey and Song (1984) and Petite et al. (1999, 2001) and references therein.

I return at this point to the explanation of what I am doing and how it connects to my broader interest in the conceptual tools developed for the analysis of long chronoseres such as Cwynar's plant particle collection, Petit's Vostok temperature records, or the Renland chronosere (Hansson 1994). Anyone contemplating the direct comparison of long natural chronoseres directly by matching time points will find it rather impossible. I devised other ways I already mentioned.

It is of interest to me to show that by taking Vostok rather than the nearby Renland as my principal temperature data source I am choosing between two snapshots of the polar climate. My modus operandi begins with the extraction of the integral function of the temperature curve from the data in high-level regression analysis then I partition it into quantitative and qualitative components. I include a refreshed version of a comparative analysis from earlier work and present a summary in Table 8 and Figure 9. I include earlier graphs and some text for convenience. Partitions of the integral equation into product components follow. The results point up the functional identity of the Vostok and Renland climate at the polar extremes in form and difference in scale.

Table 8. Regression results for the Vostok and Renland temperature chronoseres taken from Appendices 1aa and 1ba. See all graphs in Figure 8.
Vostok

$649\,^{\circ}$K readings, 0 to 41112 ice core years, 0 to 649 cm core depth
Rank 51 Eqn 7903 y=(a+cx+ex^2)/(1+bx+dx^2)
r^2 Coef Det   DF Adj r^2   Fit Std Err   F-val
0.9606627148 0.9603624302 0.6722271533 4005.0726569

Parm	Value	Std Error	t-value	90% Confidence Limits		P>|t|
a	-0.52350044	2.18656e-09	-2.3942e+08	-0.52350044	-0.52350044	0.00000
b	-0.00010575	1.02094e-06	-103.578754	-0.00010743	-0.00010407	0.00000
c	0.000162619	6.07073e-06	26.78730681	0.000152619	0.000172618	0.00000
d	3.55092e-09	7.12041e-11	49.8696644	3.43364e-09	3.66821e-09	0.00000
e	-1.3199e-08	4.90177e-10	-26.9268518	-1.4006e-08	-1.2392e-08	0.00000

Renland
Renland $811\,^{\circ}$K readings, 2040 to 41140 ice core years, 0 to 312 cm core depth
Rank 83 Eqn 7903 y=(a+cx+ex^2)/(1+bx+dx^2) [N
r^2 Coef Det   DF Adj r^2   Fit Std Err   F-val
0.8657632273 0.8649294585 2.2201997007 1299.5789955

Parm	Value	Std Error	t-value	90% Confidence Limits		P>|t|
a	-0.67531293	8.91989e-09	-7.5709e+07	-0.67531294	-0.67531291	0.00000
b	-0.00012055	1.15417e-06	-104.444467	-0.00012245	-0.00011865	0.00000
c	0.000452729	1.39798e-05	32.38443201	0.000429708	0.00047575	0.00000
d	4.34287e-09	8.45847e-11	51.34348708	4.20358e-09	4.48216e-09	0.00000
e	-3.1628e-08	1.18445e-09	-26.702944	-3.3579e-08	-2.9678e-08	0.00000

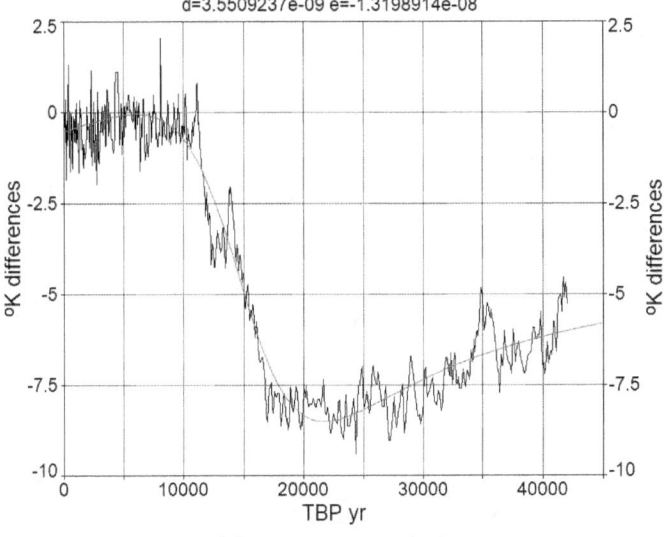

(a) Regression analysis

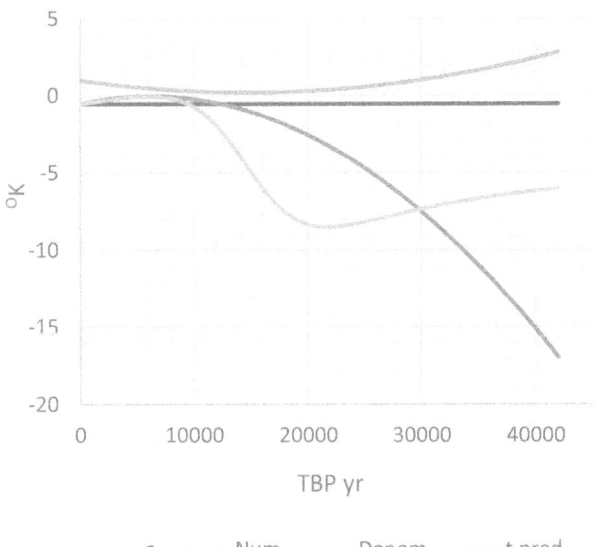

(b) Product type partition of the Vostok integral function into components, explained in the text.

Renland temperature chronosere

Rank 83 Eqn 7903 y=(a+cx+ex^2)/(1+bx+dx^2) [NL]

r^2=0.86576323 DF Adj r^2=0.86492946 FitStdErr=2.2201997 Fstat=1299.579

a=-0.67531293 b=-0.00012054678 c=0.00045272887

d=4.3428721e-09 e=-3.1628352e-08

(c) Regression analysis

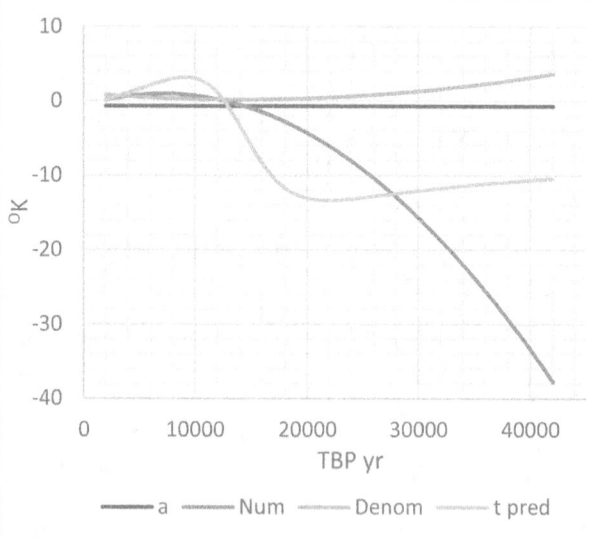

(d) Product type partition of the Renland integral function into components as explained in the text.

Figure 9. Temperature chronoseres, Vostok and Renland. Visit Table 8 for regression data. Legend: a – a regression constant, Num = a+cx+ex^2 (numerator), Denom=1+bx+dx^2 (denominator), t pred =Num/Denom (temperature prediction).

Since differences are shown on the vertical scale, the unit °K is the same as the unite °C.

What do we see in Figure 9? There are mappings of the integral functions (a,c), each factored (b,d) into shape (Num) and scale (Denom). The product Num x Denom^{-1} is a temperature prediction. In all respect, the Vostok and Renland curve-pairs are similar as if the pair were describing the polar extremes of the same global process. The observed high level of synchrony should not be unexpected to those familiar with the layout of the biota in a neatly arranged zonal continuum with a predictable shift in composition relative to the distance from the equator to and from the polar regions across latitudes depending on the climate warming or cooling cycles (Orlóci 1994, 2008). Knowing the connection of the polar regions by a dynamic pattern can be used predictively to foretell the possible consequence of climate change at either extreme. Altitudinal and other modification applies.

The following table and graphs should further clarify what I mean when I call the Vostok and Renland chronoseres similar. I give the product-moment correlation values for the functions (V Vostok, R Renland):

	Den V	t pred V	Num R	Den R	t pred R
Num V	-0.899	0.628	1.000	-0.937	0.689
Denom V		-0.278	1.000	0.996	-0.397
t pred V			0.610	0.981	0.981
Num R				-0.946	0.676
Denom R					-0.467

The correlations are based on the co-ordinates generated by regression equations. I opted to base the comparisons on the graphs of fitted functions as in Figure 1. Direct chronology-based comparisons of the temperature chronoseres were bypassed. The generic connection of Vostok and Renland is verified on the components level by the high correlation between the like components, and on the holistic level between 't pred V' and 't pred R'.

Synmechanics

Chronosere tensors

In this section, I discuss the numerical relationship of synmechanic tensors of the plant particle chronosere with concomitant global temperature oscillations. I use Cwynar's 95 palynomorph taxa and 133 paleorelevés in the analysis. The time-span is 41138 years from initiation, equivalent to the aggradation of a 398 cm deep lake sediment.

For temperature records, I had a choice between two sources, Vostok from Antarctica and Renland from Greenland. I opted for Vostok. I explain the reasons in the upcoming text.

Tensor vector analysis

The central question posed and answered in this section is: *"Do long-term atmospheric temperature oscillations affect the oscillation of the mechanical tensor elements of the vegetation process."* Having the Cwynar's plant particle chronosere and the Vostok temperature data set in hands, we have a reliable base for the examination of concomitant oscillations in the key plant particle tensors of the paleorelevés for concordance with oscillations of the ^{O}K temperature-differences in the perspective of 40+ millennia. I focus on the evolving patterns of velocity V, acceleration A, linear momentum (kinetic energy) L, force F, and work W. The relevant graphs are in Figures 1, 7, 9, 10, and 11. Figures 10, 11 have a basis for the tensor and a second graph for the residuals at lag 1. I give the graphs for V, A, L, F and W, in Figure 11. I do not give results for lag 1.

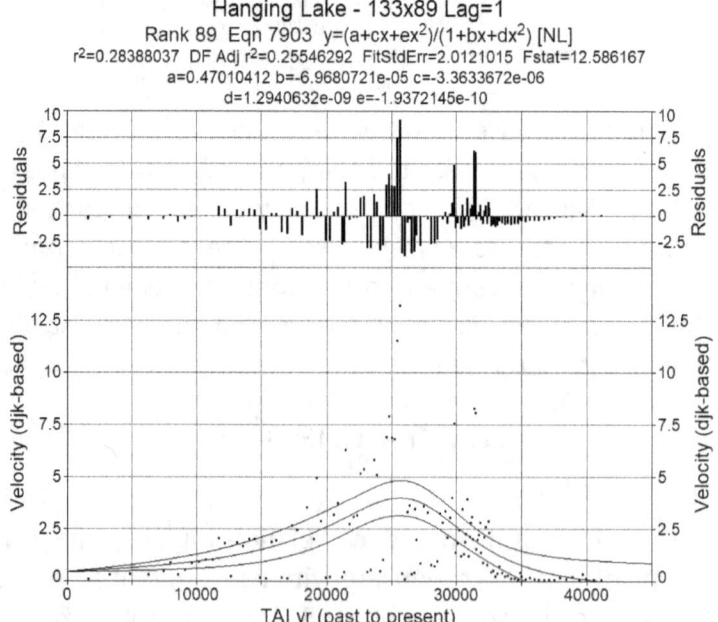

Hanging Lake - 133x89 Lag=1
Rank 89 Eqn 7903 y=(a+cx+ex²)/(1+bx+dx²) [NL]
r²=0.28388037 DF Adj r²=0.25546292 FitStdErr=2.0121015 Fstat=12.586167
a=0.47010412 b=-6.9680721e-05 c=-3.3633672e-06
d=1.2940632e-09 e=-1.9372145e-10

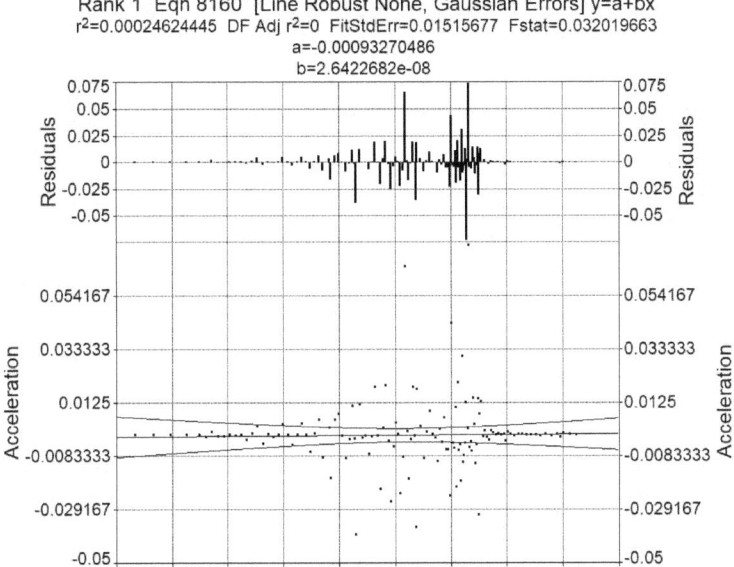

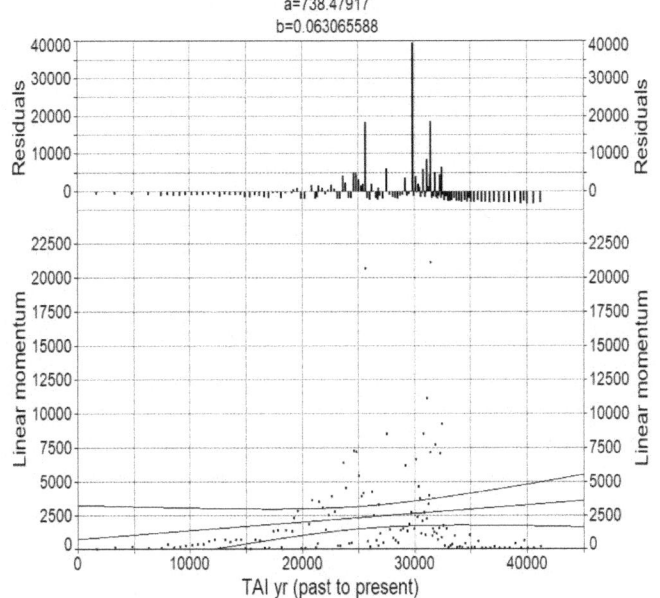

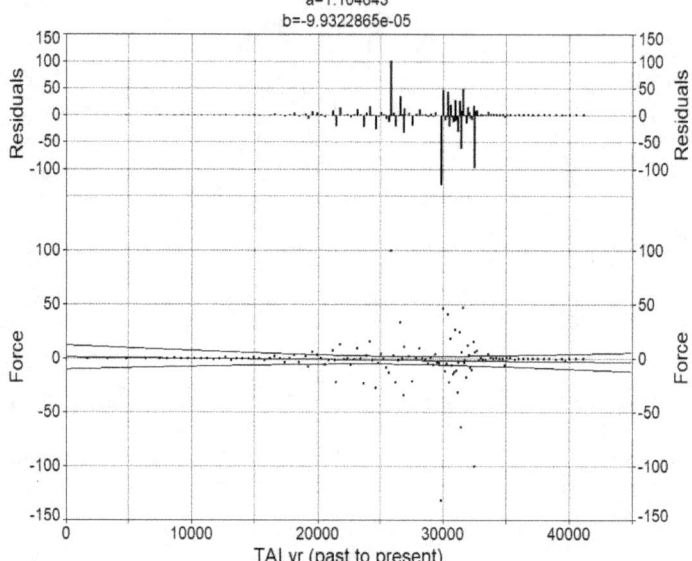

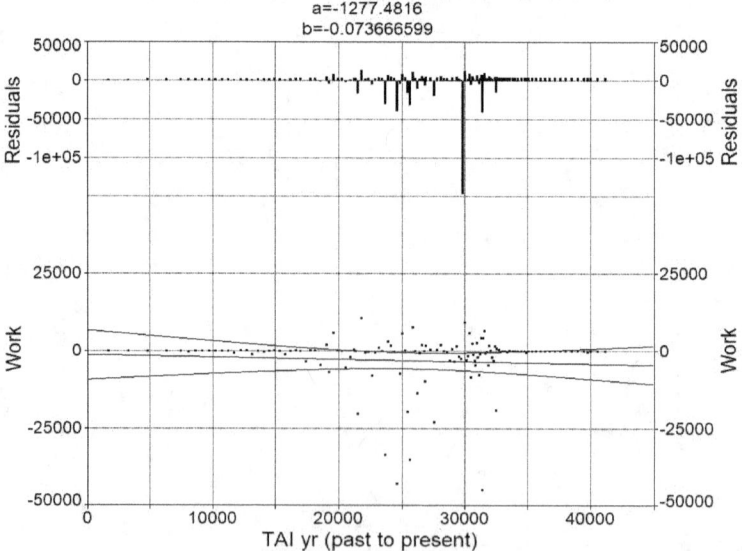

Figure 10. The relationship of plant particle composition tensors and Hanging Lake's sediment age. Zero on the time axis mark initiation time, the laying of the bottom

horizon of the sediment core where Cwynar's chronosere begins. All figures have 132 points. The analysis is based on 98 palynomorph taxa. The time-span of the record in this exercise is 41138 sediment years.

Considering Figure 7 and Appendix 7, note the substantial oscillation of temperature in the graph of residuals for the full length of the Cwynar chronosere. Note too, years 23661 and 29901 on the horizontal axis. These mark inflexion points on the temperature graph corresponding to the 2nd differential equation's maximum and minimum.

I turn to the acceleration results (Figure 10) to establish guidelines by example for the interpretation of the graphs:

1. Inspect the oscillation patterns of the residuals and make comparisons with the oscillation pattern of the Vostok temperature residuals.

2. All graphs have 132 points based on the analysis of 133 paleorelevés. The acceleration tensor vector tunnels through temperature oscillations (the residuals) on the entire length of the chronosere.

3. Note, the temperature oscillation's rather uniform periodic pattern from end to end, but the pattern of the acceleration residuals has three phases. The first phase extends from initiation almost to the maximum of temperature's 2nd differential equation (Figure 7, year 23661 TAI). At that point, acceleration enters a period of extreme oscillations with considerably irregular amplitude until the periods end just beyond the minimum of temperature's 2nd differential equation (year 29901 TAI).

3. Clearly, the onset of high A oscillations did not start in earnest until the advent of the warming cycle which lasted for almost 10 millennia with one interruption and brought Beringia out of the deepest Ice Age into the Inter Glacial Age.

4. It is clear too, once the temperature oscillations subsided, the oscillation of A subsided as well.

4. We see intense residual tensor oscillations in general not tied simply to temperature oscillations in periods where there is no definite long-term steady climate warning.

4. We can see too, that moderate warming that lasts for millennia will carve out distinct biotic zones in a march up through latitudes on terra novae left behind the melted ice on the regional scales.

4. Short term intense climate warning that we experience today will leave in ruins today's biota by the melt of the permafrost in the tundra, to regional die back in the Boreal forest, and so forth. Should the climate stabilize and gain longevity in that state, migrations of species from refugia will build anew the optimal geographic climax pattern of biotic communities with community components draw from the surviving resources in refugia.

5. Is there any message in this for us planetary residents in a new climate warming cycle? The message is rather bleak. The normalcy built into the climax pattern through many millennia in small steps, once destroyed cannot return soon enough on the human time scale. The civilization we know may not survive a hasty disintegration of its bio-geoclimatic cradle.

One more set of graphs (Figure 11) display regression results of the tensors paired with the predicted Vostok temperature differences

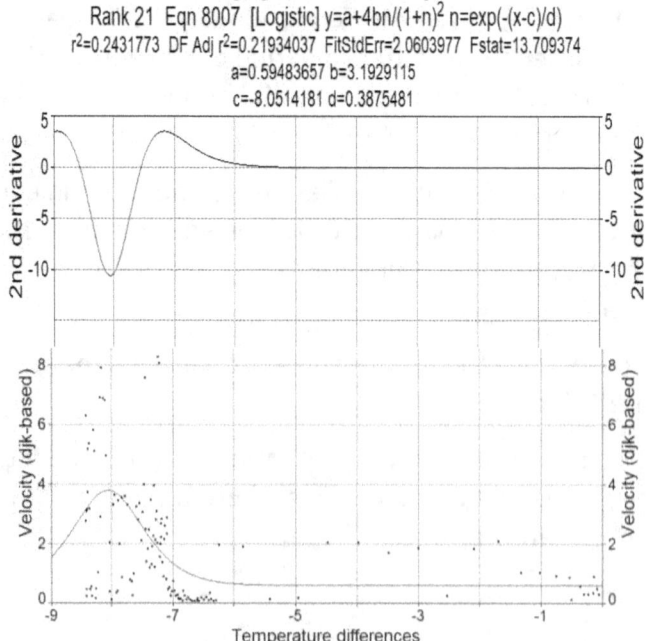

Hanging lake - 133x98 Lag=1
Rank 21 Eqn 8007 [Logistic] $y=a+4bn/(1+n)^2$ $n=exp(-(x-c)/d)$
$r^2=0.2431773$ DF Adj $r^2=0.21934037$ FitStdErr=2.0603977 Fstat=13.709374
a=0.59483657 b=3.1929115
c=-8.0514181 d=0.3875481

Hanging lake - 133x98 Lag=1
Rank 219 Eqn 1 y=a+bx
r²=0.00013120055 DF Adj r²=0 FitStdErr=0.015157643 Fstat=0.01705831
a=0.00024566812
b=7.6102476e-05

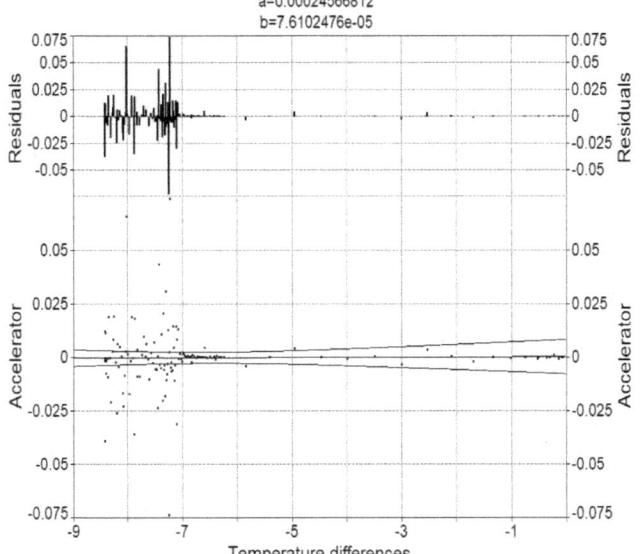

Hanging lake - 133x98 Lag=1
Rank 213 Eqn 1 y=a+bx
r²=0.043775869 DF Adj r²=0.028950689 FitStdErr=4693.4179 Fstat=5.9513903
a=-516.18416
b=-440.14682

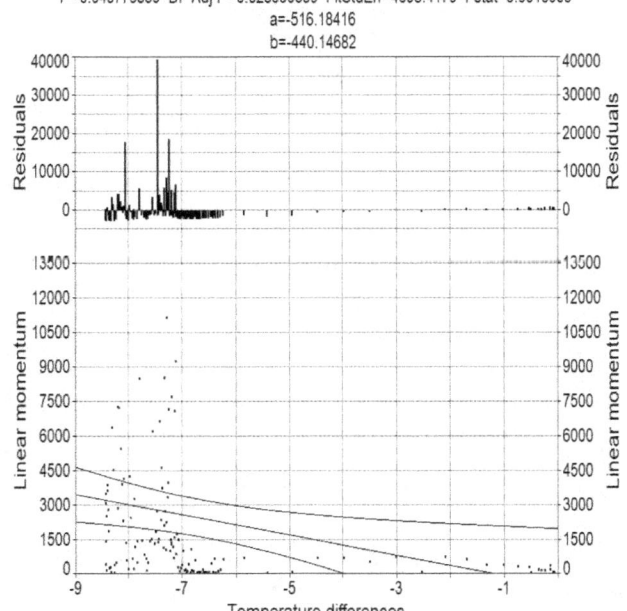

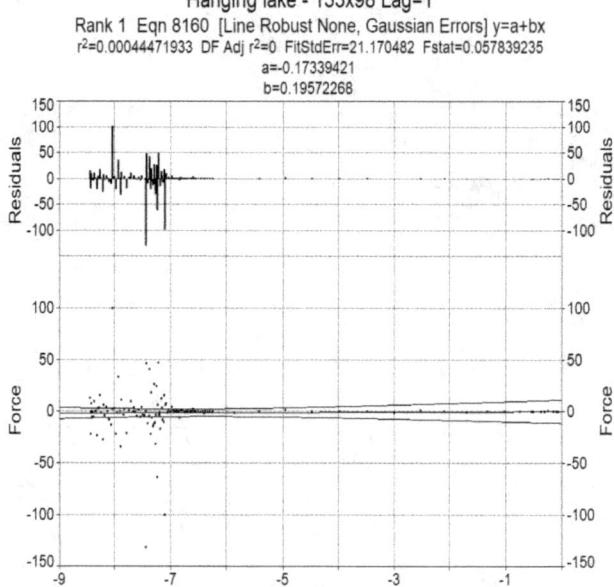

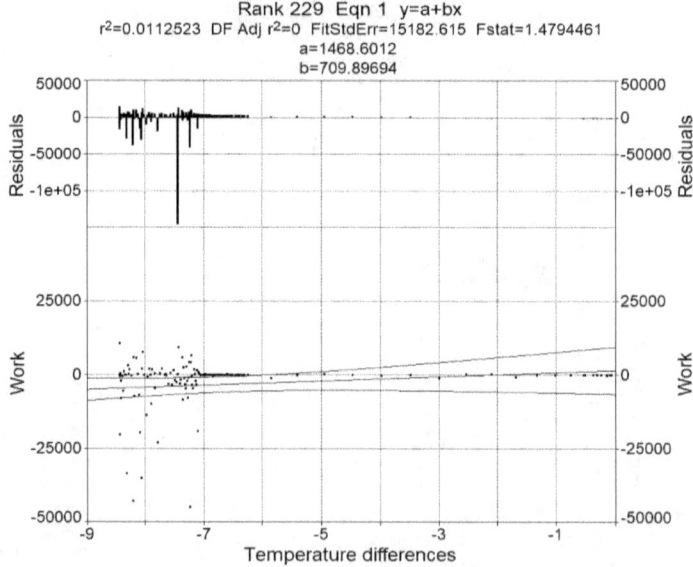

Figure 11. Predicted Vostok temperatures (see equation in Figure 6) and the states of the Cwynar chronosere's characteristic tensors. Temperature predictions by

equation 7903 of Figure 6 are the basis of the comparison. The number of points is 133 The number of palynomorph taxa in the analysis is 98.

What do we see in Figure 11? Almost uniformly the bulge in the point pattern is in the -9 to -7 temperature range. That is the beginning of the Late Quaternary warming cycle. Further explanations are left to interested readers.

Generalizations

Conceptual tools are presented for Multiscale Time Series Analysis and tested on a long chronosere of plant particle counts taken from lake sediment in Eastern Beringia's dry arctic steppe tundra. The analysis identifies statistically significant directional regularities in the plant particle chronosere from which assembly rules are inferred that control the composition of the attractor (climax) state in the source vegetation. The rules facilitate the construction of predictive assembly scenarios under an assumed change in the global climate. The results are reinforcing the idea that regularity and scale-dependence are both parts of the chaotic vegetation process; yet, random variation is dominant. Therefore, the appropriate approach to probe for regularity and scale dependence is probabilistic. Tests show increasing sharpness of regularity with increasing lag in the directional pairing of paleorelevés. This is telling us the vegetation process is indeed a multiscale phenomenon and requires a multiscale frame in its study. Periodic convergence of the vegetation process onto low-velocity phases (low values of 1st derivative) coincide with low-velocity phases in the temperature process. When the temperature's process velocity increases, the vegetation's process velocity increases too. This makes vegetation science's climax theory congruent with the attractor concept of general systems theory.

Literature

Internet addresses identify availability for resources of free download or paid distribution:

Beer, F.P. and E.R. Johnston. 1990. Vector Mechanics for Engineers. 2nd metric ed. McGraw-Hill, New York.

Cwynar, L.C. 1982. A late-Quaternary vegetation history from Hanging Lake, northern Yukon. Ecological Monographs 52: 1-24.

Cwynar, L.C. 1982. Global Pollen Database (2000).
http://www.ncdc.noaa.gov/paleo/ftp-pollen.html
http://www.lpc.uottawa.ca/data/cpd/HangingLake.xls
http://hurricane.ncdc.noaa.gov/pls/paleox/f?p=519:1:0::::P1_STUDY_ID:7483

Cwynar, L.C. and J.C. Ritchie. 1980. Artic steppe tundra: a Yukon perspective. Science 208: 1375-1377.

Gleick, 1988. Chaos. Making a new science. Penguin Books, U.S.A.

Hansson, M.E. 1994. The Renland ice core. A Northern Hemisphere record of aerosol composition over 120,000 years. Tellus 46(B), 390-418.

Hultén, E. 1937. Outline of the history of arctic and boreal biota during the Quaternary Period. Lehre J. Cramer, New York.

Kruskal, J.B. 1964. Multidimensional scaling by optimizing goodness of fit to a nonmetric hypothesis. Psychometrika 29: 1-27.

Kurek, J., L.C. Cwynar and J.C. Vermoire. 2009. A Late-Quaternary paleo temperature record from hanging Lake, northern Yukon Territory, Eastern Beringia. Quaternary Research 2: 246-2572.

Orlóci, L. 1980. Non-linear data structures and their description. In: L. Orlóci, C.R. Rao and W.M. Stiteler, Multivariate methods in ecological work, pp. 239-275. ICP, Fairland, Maryland. https://www.researchgate.net/profile/Laszlo_Orloci2

Orlóci, L. 1991. On character-based community analysis: choice, arrangement, comparison. In: Feoli, E. and L. Orlóci (eds.) Computer Assisted Vegetation Analysis, pp. 81-93. Kluwer Academic Publishers, London. https://www.researchgate.net/profile/Laszlo_Orloci2

Orlóci, L. 1994. Global warming: the process and its phytoclimatic consequences in temperate and cold zones. Coenoses 9: 69-74. https://www.researchgate.net/profile/Laszlo_Orloci2

Orlóci, L. 2008. Vegetation displacement issues and transition statistics in climate warming cycle. Community Ecology 9: 83-98. https://www.researchgate.net/profile/Laszlo_Orloci2

Orlóci, L. 2009. Multiscale trajectory analysis. Frontiers of Biology in China 4: 158-179. https://www.researchgate.net/profile/Laszlo_Orloci2

Orlóci, L. 2012. Statistical multiscaling in dynamic ecology. Probing the long-term vegetation process for patterns of parameter oscillations. Scada Publishing, Canada. Online edition: https://www.amazon.ca/dp/1475071388 https://www.researchgate.net/profile/Laszlo_Orloci2

Orlóci, L. 2014a. Quantum analysis of primary succession. The energy structure of a vegetation chronosere in Hawai'i Volcanoes National Park. SCADA Publishing, Canada. Online Edition: https://www.researchgate.net/profile/Laszlo_Orloci2 https://www.amazon.ca/dp/1492788996

Orlóci, L. 2014b. The vegetation process. A holistic study of long-term community energetics in East Beringia. Scada Publishing, Canada. Online edition: https://www.researchgate.net/profile/Laszlo_Orloci2 https://www.amazon.ca/dp/1499142064

Orlóci, L. 2015. Diversity analysis, holistic energetics, and statistics. The resonator complex model of the vegetation stand. SCADA Publishing, Canada. Online edition: https://www.researchgate.net/profile/Laszlo_Orloci2 https://www.amazon.ca/dp/1517687063

Orlóci, L. 2019. Statistical Ecology. Quantitative exploration of nature to reveal the unexpected. SCADA Publishing, Canada. Online Edition: https://www.researchgate.net/profile/Laszlo_Orloci2 https://www.amazon.ca/dp/1453760520

Orlóci, L. 2019. Looking back. A life of statistical ecology. 2nd Enlarged ed. Scada Publishing, Canada. Online Edition: https://www.amazon.com/dp/1796403148 https://www.researchgate.net/profile/Laszlo_Orloci2

Orlóci, L. 2019. Statistical quantum ecology. Essays on the resonator complex model of the vegetation stand. SCADA Publishing, Canada. Online Edition: https://www.researchgate.net/profile/Laszlo_Orloci2 https://www.amazon.ca/dp/153716788X

Orlóci, L. ResearchGate page. https://www.researchgate.net/profile/Laszlo_Orloci2

Orlóci, L., M. Anand, and X.S. He. 1993. Markov chain: a realistic model for temporal coenosere? Biométrie-Proximétrie 33: 7-26. https://www.researchgate.net/profile/Laszlo_Orloci2

Planck, M. 1901. On the law of distribution of energy in the normal spectrum. Annalen der Physik Vol. 4, p. 553 at seq. https://www.researchgate.net/profile/Laszlo_Orloci2

Polányi, M. 1968. Life's irreducible structure. Science 160, 3834: 1308-1312.

Schweingruber, F.H. 1996. Tree rings and environment dendroecology. Paul Haupman, Stuttgart.

Vinther, B.M., H.B. Clausen, D.A. Fisher, R.M. Koerner, S.J. Johnsen, K.K. Andersen, D. Dahl-Jensen, S.O. Rasmussen, J.P. Steffensen, and A.M. Svensson. 2008. Synchronizing ice cores from the Renland and Agassiz ice caps to the Greenland Ice Core Chronology. J. Geophys. Res., 113, D08115, doi:10.1029/2007JD009143

Whittaker, R.H. 1953. A consideration of climax theory: the climax as a population and pattern. Ecological Monographs 23: 41-78.

Petit, J.R., Jouzel, D. Raynaud, D., Barkov, N.I, Barnola, J.M., Basile, I., Bender, M., Chappellaz, J., Davis, J. , Delaygue, G., Delmotte, M., Kotlyakov, V.M., Legrand, M., Lipenkov, V., Lorius, C., Pepin, L., Ritz, C., Saltzmann, E., and M. Stievenard. 1999. Climate and atmospheric history of the past 420,000 years from the Vostok Ice Core, Antarctica. Nature 300: 429-436

Petit, J.R., Jouzel, J. Raynaud, D., Barkov, N.I, Barnola, J.M., Basile, I., Bender, M., Chappellaz, J., Davis, J. , Delaygue, G., Delmotte, M., Kotlyakov, V.M., Legrand, M., Lipenkov, V., Lorius, C., Pepin, L., Ritz, C., Saltzmann, E., and M. Stievenard. 2001.

Vostok Ice Core Data for 420,000 years, IGBP PAGES/World Data Centre for Paleoclimatology Data Contribution Series #2001-076. NOAA/NGDC Paleoclimatology Program, Boulder CO, USA.

Appendices

Appendix 1a

Numerics from high-level regression analysis of the Vostok temperature chronosere. Supplement to Figure 1a.

Rank 51 Eqn 7903 y=(a+cx+ex^2)/(1+bx+dx^2) [N

r^2 Coef Det	DF Adj r^2	Fit Std Err	F-val
0.9606627148	0.9603624302	0.6722271533	4005.0726569

Parm	Value	Std Error	t-value	90% Confidence Limits		P>|t|
a	-0.52350044	2.18656e-09	-2.3942e+08	-0.52350044	-0.52350044	0.00000
b	-0.00010575	1.02094e-06	-103.578754	-0.00010743	-0.00010407	0.00000
c	0.000162619	6.07073e-06	26.78730681	0.000152619	0.000172618	0.00000
d	3.55092e-09	7.12041e-11	49.8696644	3.43364e-09	3.66821e-09	0.00000
e	-1.3199e-08	4.90177e-10	-26.9268518	-1.4006e-08	-1.2392e-08	0.00000

Area Xmin-Xmax	Area Precision	
-210430.0055	3.937756e-11	

Function min	X-Value	Function max	X-Value
-8.506146147	21665.832708	-0.046453233	6049.8204988

1st Deriv min	X-Value	1st Deriv max	X-Value
-0.00108986	14547.863426	0.000170003	27023.463197

2nd Deriv min	X-Value	2nd Deriv max	X-Value
-1.94286e-07	11379.545173	2.168559e-07	17799.332237

Singularities [Data Range] None
Singularities [All Other] None

Soln Vector	Covar Matrix	SVD Cond
LvMrq/SVD	SVDecomp	1.611317e+20

r^2 Coef Det	DF Adj r^2	Fit Std E
0.9606627148	0.9603624302	0.6722271533

Source	Sum of Squares	DF	Mean Square	F Statistic	P>F
Regr	7239.3986	4	1809.8497	4005.07	0.00000
Error	296.43941	656	0.45188935		
Total	7535.8381	660			

Description: Vostok temperature chronosere
X Variable: TBP

Xmin: 0   Xmax: 42001   Xrange: 42001
Xmean: 17636.951589   Xstd: 12708.225778 Xmedian: 15099
X@Ymin: 24363   X@Ymax: 8135   X@Yrange: 16228
Variable: oK differences
Ymin: -9.39   Ymax: 2.06   Yrange: 11.45
Ymean: -4.065839637   Ystd: 3.3790437187 Ymedian: -4.68
Y@Xmin: 0   Y@Xmax: -5.27   Y@Xrange: 5.27

Date	Time	File Source
May 20, 2020	8:40:49 AM	CLIPBRD.PRN

Appendix 1b

Numerics from high-level regression analysis of the Renland temperature chronosere. Supplement to Figure 1b.

Rank 83 Eqn 7903 y=(a+cx+ex^2)/(1+bx+dx^2) [N

r^2 Coef Det	DF Adj r^2	Fit Std Err	F-val
0.8657632273	0.8649294585	2.2201997007	1299.5789955

Parm Value   Std Error   t-value   90% Confidence Limits   P>\|t\|

a   -0.67531293 8.91989e-09 -7.5709e+07 -0.67531294 -0.67531291 0.00000
b   -0.00012055 1.15417e-06 -104.444467 -0.00012245 -0.00011865 0.00000
c   0.000452729 1.39798e-05 32.38443201 0.000429708 0.00047575 0.00000
d   4.34287e-09 8.45847e-11 51.34348708 4.20358e-09 4.48216e-09 0.00000
e   -3.1628e-08 1.18445e-09 -26.702944 -3.3579e-08 -2.9678e-08 0.00000
Area Xmin-Xmax Area Precision
-302167.9097 5.818422e-10
Function min X-Value   Function max X-Value
-13.38684935 22055.038608 3.143609177 9183.6242726
1st Deriv min X-Value   1st Deriv max X-Value
-0.002663214 14420.918897 0.0005103692 5132.7017028
2nd Deriv min X-Value   2nd Deriv max X-Value
-7.06831e-07 11800.496803 5.675607e-07 16909.45552
Singularities [Data Range] None
Singularities [All Other] None
Soln Vector   Covar Matrix   SVD Cond
LvMrq/SVD   SVDecomp   2.164527e+20
r^2 Coef Det   DF Adj r^2   Fit Std E
0.8657632273 0.8649294585 2.2201997007
Source Sum of Squares DF   Mean Square   F Statistic   P>F
Regr   25623.99   4   6405.9975   1299.58   0.00000
Error   3973.0051   806   4.9292867
Total   29596.995   810
Description: Renland temperature chronosere
X Variable: TBP yr
   Xmin: 2040   Xmax: 42070   Xrange: 40030
   Xmean: 12419.066584   Xstd: 9069.7371687 Xmedian: 10140
   X@Ymin: 27590   X@Ymax: 10820   X@Yrange: 16770
Y Variable: oK differences
   Ymin: -16.24242424   Ymax: 6.484848485 Yrange: 22.727272725
   Ymean: -1.992003886 Ystd: 6.0447911441 Ymedian: 0.484848485
   Y@Xmin: 0   Y@Xmax: -10.66666667 Y@Xrange: 10.66666667
Date   Time   File Source
May 21, 2020 7:46:28 AM   CLIPBRD.PRN

Appendix 3a

Rank 2185 Eqn 8160 [Line Robust None, Gaussian Errors] y=a+bx
r^2 Coef Det   DF Adj r^2   Fit Std Err   F-val
0.9301924455 0.9219797921 0.0127714145 239.85174878
Parm   Value   Std Error   t-value   90% Confidence Limits   P>|t|
a   0.985875789 0.005932724 166.1759134 0.975588069 0.99616351 0.00000
b   -0.00767008 0.000495254 -15.4871479 -0.00852888 -0.00681127 0.00000
Procedure   Minimization Iterations
LevMarqdt   Least Squares 6
r^2 Coef Det   DF Adj r^2   Fit Std E
0.9301924455 0.9219797921 0.0127714145
Source Sum of Squares DF   Mean Square   F Statistic   P>F
Regr   0.039121986   1   0.039121986   239.852   0.00000
Error 0.0029359625 18   0.00016310903
Total 0.042057948   19
May 6, 2020 12:25:27 PM CLIPBRD.PRN

Appendix 3b

Rank 2185 Eqn 8160 [Line Robust None, Gaussian Errors] y=a+bx
XY X Value Y Value Y Predict Residual Residual% 90% Confidence Limits 90% Prediction Limits

1	1	0.9847	0.9782057	0.0064943	0.6595192	0.9686618	0.9877496	0.9540903	1.0023211
2	2	0.9744	0.9705356	0.0038644	0.3965888	0.9617146	0.9793567	0.9466971	0.9943742
3	3	0.9712	0.9628656	0.0083344	0.8581586	0.9547409	0.9709902	0.9392758	0.9864553
4	4	0.9493	0.9551955	-0.005895	-0.621035	0.9477333	0.9626577	0.9318256	0.9785653
5	5	0.9499	0.9475254	0.0023746	0.2499828	0.9406819	0.9543689	0.9243457	0.9707051
6	6	0.9385	0.9398553	-0.001355	-0.144415	0.9335737	0.9461369	0.9168353	0.9628754
7	7	0.9419	0.9321853	0.0097147	1.0313979	0.9263923	0.9379782	0.9092937	0.9550768
8	8	0.9147	0.9245152	-0.009815	-1.07305	0.9191177	0.9299127	0.9017205	0.9473099
9	9	0.9138	0.9168451	-0.003045	-0.333236	0.9117282	0.921962	0.8941152	0.939575
10	10	0.9075	0.909175	-0.001675	-0.184577	0.9042044	0.9141457	0.8864776	0.9318725
11	11	0.9139	0.901505	0.012395	1.3562794	0.8965343	0.9064756	0.8788075	0.9242024
12	12	0.8839	0.8938349	-0.009935	-1.123983	0.888718	0.8989518	0.871105	0.9165648
13	13	0.854	0.8861648	-0.032165	-3.766371	0.8807673	0.8915623	0.8633701	0.9089595
14	14	0.8836	0.8784947	0.0051053	0.5777799	0.8727018	0.8842877	0.8556032	0.901386
15	15	0.8662	0.8708247	-0.004625	-0.533902	0.8645431	0.8771063	0.8478046	0.8938447
16	16	0.864	0.8631546	0.0008454	0.0978488	0.8563111	0.8699981	0.8399749	0.8863343
17	17	0.8328	0.8554845	-0.022685	-2.723885	0.8480223	0.8629467	0.8321147	0.8788544
18	18	0.8624	0.8478144	0.0145856	1.691276	0.8396898	0.8559391	0.8242247	0.8714042
19	19	0.846	0.8401444	0.0058556	0.6921559	0.8313233	0.8489654	0.8163058	0.8639829
20	20	0.8541	0.8324743	0.0216257	2.5319886	0.8229304	0.8420182	0.8083589	0.856589

Appendix 4a

Rank 3142 Eqn 8160 [Line Robust None, Gaussian Errors] y=a+bx

r^2 Coef Det	DF Adj r^2	Fit Std Err	F-value	
0.879733696	0.865584719	8.018272612	131.6678573	

Parm	Value	Std Error	t-value	90% Confidence Limits		P>|t|
a	-360.820494	35.44254452	-10.1804342	-422.280122	99.36086	0
b	448.6386259	39.09820126	11.47466153	380.8398581	516.437	0

Procedure	Minimization	Iterations
LevMarqdt	Least Squares	8

r^2 Coef Det	DF Adj r^2	Fit Std Err
0.879733696	0.865584719	8.018272612

Source	Sum of Squares	DF	Mean Square	F Statistic	P>F
Regr	8465.2815	1	8465.2815	131.668	0
Error	1157.2685	18	64.292696		
Total	9622.55	19			

Date	Time	File Source
May 6, 2020	0.722858796	CLIPBRD.PRN

Appendix 4b

Rank 3142 Eqn 8160 [Line Robust None, Gaussian Errors] y=a+bx

XY	X Value	Y Value	Y Predict	Residual	Residual%	90% Confidence Limits		90% Prediction limits	
1	0.8328	18	12.805754	5.1942459	28.856922	6.9873126	18.624196	-2.266767	27.878275
2	0.846	25	18.727784	6.2722161	25.088864	13.643268	23.8123	3.9230911	33.532477
3	0.854	22	22.316893	-0.316893	-1.440422	17.64975	26.984036	7.6503033	36.983483
4	0.8541	19	22.361757	-3.361757	-17.69346	17.699668	27.023845	7.6967747	37.026739
5	0.8624	17	26.085457	-9.085457	-53.44387	21.826129	30.344786	11.543501	40.627414
6	0.864	21	26.803279	-5.803279	-27.63466	22.617347	30.989211	12.282651	41.323908
7	0.8662	25	27.790284	-2.790284	-11.16114	23.702723	31.877845	13.297707	42.282861
8	0.8836	21	35.596596	-14.5966	-69.5076	32.155834	39.037359	21.272998	49.920195
9	0.8839	44	35.731188	8.2688121	18.792755	32.299089	39.163286	21.409668	50.052708
10	0.9075	54	46.319059	7.6809406	14.223964	43.20654	49.431579	32.070747	60.567372
11	0.9138	47	49.145483	-2.145483	-4.564857	45.983945	52.307021	34.886382	63.404583
12	0.9139	40	49.190347	-9.190347	-22.97587	46.027572	52.353122	34.930972	63.449722
13	0.9147	66	49.549258	16.450742	24.925367	46.376082	52.722433	35.287572	63.810943
14	0.9385	66	60.226857	5.7731432	8.7471866	56.390092	64.063622	45.803009	74.650705
15	0.9419	63	61.752228	1.2477718	1.9805902	57.775999	65.728457	47.290657	76.213799

16	0.9493	76	65.072154	10.927846	14.378745	60.765262	69.379046	50.516195	79.628113
17	0.9499	67	65.341337	1.6586628	2.4756162	61.006195	69.676479	50.776994	79.90568
18	0.9712	70	74.89734	-4.89734	-6.9962	69.456331	80.338349	59.966459	89.828221
19	0.9744	73	76.332984	-3.332984	-4.565731	70.712559	81.953408	61.33579	91.330177
20	0.9847	73	80.953961	-7.953961	-10.89584	74.739766	87.168157	65.724291	96.183631

Appendix 5a

Rank 285 Eqn 2040 y=a+bx+cx^2+dx

r^2 Coef Det	DF Adj r^2	Fit Std Err	F-val
0.9707363744	0.9638508154	1465.5786384	199.03269416

Parm	Value	Std Error	t-value	90% Confidence Limits		P>|t|
a	537.8332017	1162.598535	0.462613005	-1478.1866	2553.853008	0.64919
b	126.3930639	25.12278804	5.03101263	82.82855155	169.9575763	0.00009
c	-0.62937589	0.143742296	-4.37850175	-0.87863418	-0.38011761	0.00036
d	0.001365206	0.000235887	5.787552007	0.000956164	0.001774248	0.00002

Area Xmin-Xmax	Area Precision
5545008.0973	0

Function min	X-Value	Function max	X-Value
2077.5801326	13.000020482	37215.746024	398

1st Deriv min	**X-Value**	**1st Deriv max**	**X-Value**
29.676525425	**153.67054829**	**274.17215699**	**398**

2nd Deriv min	X-Value	2nd Deriv max	X-Value
-1.152265545	13.000020482	2.0013602934	398

Soln Vector	Covar Matrix
GaussElim	LUDecomp

r^2 Coef Det	DF Adj r^2	Fit Std E
0.9707363744	0.9638508154	1465.5786384

Source	Sum of Squares	DF	Mean Square	F Statistic	P>F
Regr	1.2825194e+09	3	4.2750645e+08	199.033	0.00000
Error	38662573	18	2147920.7		
Total	1.3211819e+09	21			

Date	Time	File Source
May 7, 2020	10:54:58 AM	CLIPBRD.PRN

Appendix 5b

Rank 285 Eqn 2040 y=a+bx+cx^2+dx

XY *	X Value	Y Value	Y Predict	Residual	Residual%	90% Confidence Limits		90% Prediction Limits	
1	13	1870	2077.5779	-207.5779	-11.10042	496.01622	3659.1395	-915.7622	5070.9179
2	23	2930	3128.5443	-198.5443	-6.776256	1808.083	4449.0056	264.5671	5992.5215
3	28	5100	3613.3773	1486.6227	29.149465	2398.8866	4827.868	796.68864	6430.066
4	53	4360	5671.9965	-1311.996	-30.09166	4748.4663	6595.5267	2967.9892	8376.0038
5	78	5640	7215.2305	-1575.231	-27.92962	6307.3696	8123.0914	4516.5348	9913.9263
6	103	7870	8371.0675	-501.0675	-6.366804	7420.7994	9321.3355	5657.8117	1.108e+04
7	115.5	1.02e+04	8843.7083	1356.2917	13.296977	7886.199	9801.2176	6127.908	1.156e+04
8	128	1.15e+04	9267.4954	2232.5046	19.413084	8316.6034	1.022e+04	6554.0211	1.198e+04
9	140.5	9220	9658.4272	-438.4272	-4.755176	8726.7693	1.059e+04	6951.6332	1.237e+04
10	153	1.05e+04	1.003e+04	467.49765	4.4523586	9129.4684	1.094e+04	7335.4266	1.273e+04
11	178	1.09e+04	1.079e+04	105.92358	0.9717759	9958.1475	1.163e+04	8118.7214	1.347e+04
12	203	1.17e+04	1.168e+04	19.794321	0.1691822	1.089e+04	1.247e+04	9019.3981	1.434e+04
13	228	1.26e+04	1.282e+04	-218.8782	-1.737128	1.203e+04	1.36e+04	1.016e+04	1.548e+04
14	253	1.24e+04	1.434e+04	-1938.082	-15.62969	1.352e+04	1.516e+04	1.167e+04	1.701e+04
15	263	1.28e+04	1.508e+04	-2280.982	-17.82017	1.425e+04	1.592e+04	1.241e+04	1.776e+04
16	273	1.58e+04	1.591e+04	-113.4356	-0.717947	1.506e+04	1.677e+04	1.323e+04	1.859e+04
17	278	1.77e+04	1.637e+04	1334.1948	7.5378236	1.551e+04	1.723e+04	1.368e+04	1.905e+04
18	293	1.68e+04	1.788e+04	-1079.771	-6.427211	1.699e+04	1.877e+04	1.519e+04	2.057e+04
19	303	2.02e+04	1.903e+04	1169.9641	5.7919016	1.813e+04	1.993e+04	1.633e+04	2.173e+04
20	318	2.02e+04	2.099e+04	-787.3409	-3.897727	2.004e+04	2.193e+04	1.828e+04	2.37e+04
21	323	2.49e+04	2.171e+04	3194.2862	12.828459	2.074e+04	2.267e+04	1.899e+04	2.442e+04
22	398	3.65e+04	3.722e+04	-715.746	-1.960948	3.48e+04	3.963e+04	3.371e+04	4.072e+04

Appendix 6aa

Rank 103   Eqn 6057   High Precision Polynomial Order 7

r^2 Coef Det	DF Adj r^2	Fit Std Err	F-val
0.9999995977	0.9999987931	0.0409100034	1775516.108

Parm	Value	Std Error	t-value	90% Confidence Limits		P>|t|
a	33.43599154	0.512783496	65.20489012	32.40270799	34.46927509	0.00000
b	242.2062326	0.973181074	248.8809524	240.2452257	244.1672395	0.00000
c	-81.7945565	0.658956659	-124.127369	-83.1223861	-80.466727	0.00000
d	15.68919816	0.216037113	72.62269877	15.25387293	16.12452339	0.00000
e	-1.82240947	0.038186796	-47.7235499	-1.89935771	-1.74546123	0.00000
f	0.126990731	0.003717522	34.1600491	0.119499744	0.134481717	0.00000
g	-0.00488436	0.000187425	-26.0603458	-0.00526203	-0.00450669	0.00000
h	7.96773e-05	3.81901e-06	20.8633352	7.19818e-05	8.73728e-05	0.00000

Area Xmin-Xmax	Area Precision
4065.216007	0

Function min	X-Value	Function max	X-Value
207.83684331	1.0000016886	352.71975865	11.221316031
1st Deriv min	X-Value	1st Deriv max	X-Value
-0.132343612	12.074862592	119.00113125	1.0000016886
2nd Deriv min	X-Value	2nd Deriv max	X-Value
-88.92611042	1.0000016886	1.7398926628	13

Soln Vector	Covar Matrix
GaussElim	LUDecomp

r^2 Coef Det	DF Adj r^2	Fit Std E
0.9999995977	0.9999987931	0.0409100034

Source	Sum of Squares	DF	Mean Square	F Statistic	P>F
Regr	20800.879	7	2971.5541	1.77552e+06	0.00000
Error	0.0083681419	5	0.0016736284		
Total	20800.887	12			

Description: #,it100 11x65

X Variable: Iteration

Xmin: 1	Xmax: 13	Xrange: 12
Xmean: 7	Xstd: 3.8944404818	Xmedian: 7
X@Ymin: 1	X@Ymax: 13	X@Yrange: 12

Y Variable: Sum % diff

Ymin: 207.83565	Ymax: 352.68958	Yrange: 144.85393
Ymean: 333.51381385	Ystd: 41.634208041	Ymedian: 352.014
Y@Xmin: 207.83565	Y@Xmax: 352.68958	Y@Xrange: 144.85393

Date	Time	File Source
May 13, 2020	11:31:17 AM	CLIPBRD.PRN

Appendix 6ab

Rank 103   Eqn 6057   High Precision Polynomial Order 7

XY	*	X Value	Y Value	Y Predict	Residual	Residual%	90% Confidence Limits		90% Prediction Limits	
1	!	1	207.83565	207.83664	-0.000992	-0.000477	207.75437	207.91892	207.72018	207.95311
2		2	290.78777	290.78657	0.0012025	0.0004135	290.70897	290.86416	290.67336	290.89978
3		3	327.38243	327.36917	0.0132628	0.0040512	327.30877	327.42957	327.26697	327.47136
4		4	342.41582	342.45748	-0.041657	-0.012166	342.39895	342.51601	342.35638	342.55858
5		5	348.53451	348.49978	0.0347264	0.0099635	348.44602	348.55355	348.40137	348.5982
6		6	351.01265	350.99323	0.0194248	0.0055339	350.93756	351.04889	350.89375	351.0927
7		7	352.014	352.04694	-0.032941	-0.009358	351.99539	352.09849	351.94971	352.14417
8		8	352.41833	352.43635	-0.018025	-0.005115	352.38069	352.49202	352.33688	352.53583
9		9	352.58164	352.55015	0.0314888	0.0089309	352.49639	352.60391	352.45173	352.64857
10		10	352.64765	352.63154	0.0161119	0.0045688	352.57301	352.69007	352.53044	352.73264

11	11	352.67436	352.71535	-0.040986	-0.011622	352.65494	352.77575	352.61315	352.81754
12	12	352.68519	352.66256	0.0226317	0.006417	352.58496	352.74015	352.54935	352.77577
13	13	352.68958	352.69383	-0.004247	-0.001204	352.61155	352.7761	352.57736	352.81029

Appendix 6ba

Rank 41   Eqn 6855   Chebyshev=>Std Polynomial Order 5

r^2 Coef Det	DF Adj r^2	Fit Std Err	F-val
0.9999995266	0.9999966861	0.1372284702	844913.61013

Parm	Value	Std Error	t-value	90% Confidence Limits	P>|t|
a	2164.59389				
b	372.1047804				
c	-98.5329077				
d	13.53019868				
e	-0.95235316				
f	0.027240831				

Area Xmin-Xmax	Area Precision		
18890.574432	0		
Function min	X-Value	Function max	X-Value
2450.7711839	1.0000015787	2754.5767735	8
1st Deriv min	X-Value	1st Deriv max	X-Value
0.8493520065	8	211.95615243	1.0000015787
2nd Deriv min	X-Value	2nd Deriv max	X-Value
-126.767951	1.0000015787	-0.077353874	8

Soln Vector	Covar Matrix		
GaussElim	LUDecomp		
r^2 Coef Det	DF Adj r^2	Fit Std E	
0.9999995266	0.9999966861	0.1372284702	
Source	Sum of Squares	DF	Mean Square	F Statistic	P>F
Regr	79555.6	5	15911.12	844914	0.00000
Error	0.037663306	2	0.018831653		
Total	79555.637	7			

Description: #,Sum ABS diffs %
X Variable: Recursion (present to future)

	Xmin: 1	Xmax: 8	Xrange: 7
	Xmean: 4.5	Xstd: 2.4494897428	Xmedian: 4.5
	X@Ymin: 1	X@Ymax: 8	X@Yrange: 7

Y Variable: Sum diffs %

	Ymin: 2450.782675	Ymax: 2754.584997	Yrange: 303.802322
	Ymean: 2684.4668994	Ystd: 106.60718109	Ymedian: 2734.6939835
	Y@Xmin: 2450.782675	Y@Xmax: 2754.584997	Y@Xrange: 303.802322

Date	Time	File Source
May 14, 2020	10:27:11 PM	CLIPBRD.PRN

Appendix 6bb

Rank 41   Eqn 6855   Chebyshev=>Std Polynomial Order 5

XY *	X Value	Y Value	Y Predict	Residual	Residual%
1	1	2450.7827	2450.7708	0.0118257	0.0004825
2	2	2608.4847	2608.5475	-0.062731	-0.002405
3	3	2689.0344	2688.9063	0.1280445	0.0047617
4	4	2726.3982	2726.5114	-0.113161	-0.004151
5	5	2742.9897	2742.9768	0.0129151	0.0004708
6	6	2750.1882	2750.1358	0.0523986	0.0019053
7	7	2753.2722	2753.3097	-0.037515	-0.001363
8	8	2754.585	2754.5768	0.0082235	0.0002985

Appendix 7

Rank 57   Eqn 7903   y=(a+cx+ex^2)/(1+bx+dx^2) [N
r^2 Coef Det   DF Adj r^2   Fit Std Err   F-val

0.9627887085 0.9624993517 0.659258122 4165.6437065

Parm	Value	Std Error	t-value	90% Confidence Limits		P>|t|
a	-6.29666721	2.75733e-10	-2.2836e+10	-6.29666721	-6.29666721	0.00000
b	-6.9389e-05	2.88171e-07	-240.789924	-6.9863e-05	-6.8914e-05	0.00000
c	0.000365505	1.90448e-06	191.9188653	0.000362368	0.000368642	0.00000
d	1.29947e-09	1.1533e-11	112.6736809	1.28047e-09	1.31846e-09	0.00000
e	-5.3116e-09	5.58885e-11	-95.0400124	-5.4037e-09	-5.2196e-09	0.00000

Area Xmin-Xmax Area Precision
-207190.5398 2.733529e-11

Function min X-Value   Function max X-Value
-8.430588067 19445.106312 -0.058395717 34517.741486

1st Deriv min X-Value   1st Deriv max X-Value
-0.00014815 14023.343082 0.0011114328 26793.029444

2nd Deriv min X-Value   2nd Deriv max X-Value
-2.11732e-07 29901.373403 2.184131e-07 23661.805812

Singularities [Data Range] None
Singularities [All Other] None

Soln Vector Covar Matrix SVD Cond
LvMrq/SVD   SVDecomp   1.976047e+21

r^2 Coef Det DF Adj r^2   Fit Std E
0.9627887085 0.9624993517 0.659258122

Source	Sum of Squares	DF	Mean Square	F Statistic	P>F
Regr	7241.9095	4	1810.4774	4165.64	0.00000
Error	279.8961	644	0.43462127		
Total	7521.8056	648			

Description: Vostok temperature chronosere
X Variable: TAI (past left, present right)
   Xmin: 26   Xmax: 41138   Xrange: 41112
   Xmean: 23944.097072 Xstd: 12395.943874 Xmedian: 26425
   X@Ymin: 16775   X@Ymax: 33003   X@Yrange: 16228
Y Variable: Temperature difference oK
   Ymin: -9.39   Ymax: 2.06   Yrange: 11.45
   Ymean: -4.047103236 Ystd: 3.4070110978 Ymedian: -4.28
   Y@Xmin: -6.59   Y@Xmax: 0   Y@Xrange: 6.59
Date   Time   File Source
Apr 9, 2020 10:16:51 PM CLIPBRD.PRN

Appendix 10

Rank 2436 Eqn 8160 [Line Robust None, Gaussian Errors] y=a+bx
r^2 Coef Det DF Adj r^2   Fit Std Err F-val
0.0002462444 0   0.0151567705 0.0320196631

Parm	Value	Std Error	t-value	90% Confidence Limits		P>|t|
a	-0.0009327	0.004025664	-0.23168968	-0.00760186	0.00573645	0.81714
b	2.64227e-08	1.47662e-07	0.178940389	-2.182e-07	2.71048e-07	0.85826

Area Xmin-Xmax Area Precision
-14.48105249 5.989632e-20

Function min X-Value   Function max X-Value
-0.000888288 1681.0017505 0.0001542714 41138

1st Deriv min X-Value   1st Deriv max X-Value
2.642268e-08 37150.044616 2.642268e-08 2879.9513551

2nd Deriv min X-Value   2nd Deriv max X-Value
-1.70021e-20 40757.57837 8.501046e-21 9284.5743656

Procedure   Minimization Iterations
LevMarqdt   Least Squares 7

r^2 Coef Det   DF Adj r^2   Fit Std E
0.0002462444   0   0.0151567705

Source	Sum of Squares	DF	Mean Square	F Statistic	P>F
Regr	7.3558033e-06	1	7.3558033e-06	0.0320197	0.85826
Error	0.0298646	130	0.00022972769		
Total	0.029871956	131			

Description: Hanging Lake - 133x89 lag 1
X Variable: TAI yr (past to present)
   Xmin: 1681   Xmax: 41138   Xrange: 39457
   Xmean: 25757.272727   Xstd: 8968.1425652 Xmedian: 27663
   X@Ymin: 31393   X@Ymax: 31566   X@Yrange: 173
Y Variable: Acceleration
   Ymin: -0.073587322   Ymax: 0.073885997   Yrange: 0.147473319
   Ymean: -0.000252129   Ystd: 0.0151006687 Ymedian: -9.47858e-05
   Y@Xmin: -7.03737e-05 Y@Xmax: -5.29681e-05 Y@Xrange: 1.74056e-05

Date	Time	File Source
Apr 12, 2020	5:42:21 PM	CLIPBRD.PRN

The Methuselah pygmy Eastern White Cedar (Thuja occidentalis) of Tobermory on the Bruce Peninsula, Ontario.

www.ingramcontent.com/pod-product-compliance
Lightning Source LLC
Chambersburg PA
CBHW070513220526
45467CB00002B/647